# CONSUMER DATA PRIVACY IN A NETWORKED WORLD:

## A FRAMEWORK FOR PROTECTING PRIVACY AND PROMOTING INNOVATION IN THE GLOBAL DIGITAL ECONOMY

FEBRUARY 2012

THE WHITE HOUSE
WASHINGTON

## THE WHITE HOUSE
### WASHINGTON

February 23, 2012

Americans have always cherished our privacy. From the birth of our republic, we assured ourselves protection against unlawful intrusion into our homes and our personal papers. At the same time, we set up a postal system to enable citizens all over the new nation to engage in commerce and political discourse. Soon after, Congress made it a crime to invade the privacy of the mails. And later we extended privacy protections to new modes of communications such as the telephone, the computer, and eventually email.

Justice Brandeis taught us that privacy is the "right to be let alone," but we also know that privacy is about much more than just solitude or secrecy. Citizens who feel protected from misuse of their personal information feel free to engage in commerce, to participate in the political process, or to seek needed health care. This is why we have laws that protect financial privacy and health privacy, and that protect consumers against unfair and deceptive uses of their information. This is why the Supreme Court has protected anonymous political speech, the same right exercised by the pamphleteers of the early Republic and today's bloggers.

Never has privacy been more important than today, in the age of the Internet, the World Wide Web and smart phones. In just the last decade, the Internet has enabled a renewal of direct political engagement by citizens around the globe and an explosion of commerce and innovation creating jobs of the future. Much of this innovation is enabled by novel uses of personal information. So, it is incumbent on us to do what we have done throughout history: apply our timeless privacy values to the new technologies and circumstances of our times.

I am pleased to present this new Consumer Privacy Bill of Rights as a blueprint for privacy in the information age. These rights give consumers clear guidance on what they should expect from those who handle their personal information, and set expectations for companies that use personal data. I call on these companies to begin immediately working with privacy advocates, consumer protection enforcement agencies, and others to implement these principles in enforceable codes of conduct. My Administration will work to advance these principles and work with Congress to put them into law. With this Consumer Privacy Bill of Rights, we offer to the world a dynamic model of how to offer strong privacy protection and enable ongoing innovation in new information technologies.

One thing should be clear, even though we live in a world in which we share personal information more freely than in the past, we must reject the conclusion that privacy is an outmoded value. It has been at the heart of our democracy from its inception, and we need it now more than ever.

# Foreword

Trust is essential to maintaining the social and economic benefits that networked technologies bring to the United States and the rest of the world. With the confidence that companies will handle information about them fairly and responsibly, consumers have turned to the Internet to express their creativity, join political movements, form and maintain friendships, and engage in commerce. The Internet's global connectivity means that a single innovator's idea can grow rapidly into a product or service that becomes a daily necessity for hundreds of millions of consumers. American companies lead the way in providing these technologies, and the United States benefits through job creation and economic growth as a result. Our continuing leadership in this area depends on American companies' ability to earn and maintain the trust of consumers in a global marketplace.

Privacy protections are critical to maintaining consumer trust in networked technologies. When consumers provide information about themselves—whether it is in the context of an online social network that is open to public view or a transaction involving sensitive personal data—they reasonably expect companies to use this information in ways that are consistent with the surrounding context. Many companies live up to these expectations, but some do not. Neither consumers nor companies have a clear set of ground rules to apply in the commercial arena. As a result, it is difficult today for consumers to assess whether a company's privacy practices warrant their trust.

The consumer data privacy framework in the United States is, in fact, strong. This framework rests on fundamental privacy values, flexible and adaptable common law protections and consumer protection statutes, Federal Trade Commission (FTC) enforcement, and policy development that involves a broad array of stakeholders. This framework has encouraged not only social and economic innovations based on the Internet but also vibrant discussions of how to protect privacy in a networked society involving civil society, industry, academia, and the government. The current framework, however, lacks two elements: a clear statement of basic privacy principles that apply to the commercial world, and a sustained commitment of all stakeholders to address consumer data privacy issues as they arise from advances in technologies and business models.

To address these issues, the Administration offers *Consumer Data Privacy in a Networked World*. At the center of this framework is a Consumer Privacy Bill of Rights, which embraces privacy principles recognized throughout the world and adapts them to the dynamic environment of the commercial Internet. The Administration has called for Congress to pass legislation that applies the Consumer Privacy Bill of Rights to commercial sectors that are not subject to existing Federal data privacy laws. The Federal Government will play a role in convening discussions among stakeholders—companies, privacy and consumer advocates, international partners, State Attorneys General, Federal criminal and civil law enforcement representatives, and academics—who will then develop codes of conduct that implement the Consumer Privacy Bill of Rights. Such practices, when publicly and affirmatively adopted by companies subject to Federal Trade Commission jurisdiction, will be legally enforceable by the FTC. The United States will engage with our international partners to create greater interoperability among our respective privacy frameworks. This will provide more consistent protections for consumers and lower compliance burdens for companies.

Of course, this framework is just a beginning. Starting now, the Administration will work with and encourage stakeholders, including the private sector, to implement the Consumer Privacy Bill of Rights. The Administration will also work with Congress to write these flexible, general principles into law. The Administration is ready to do its part as a convener to achieve privacy protections that preserve consumer trust and promote innovation.

# Table of Contents

# Executive Summary

Strong consumer data privacy protections are essential to maintaining consumers' trust in the technologies and companies that drive the digital economy. The existing framework in the United States effectively addresses some privacy issues in our increasingly networked society, but additional protections are necessary to preserve consumer trust. The framework set forth in this document will provide these protections while promoting innovation.

The Administration's framework consists of four key elements: A Consumer Privacy Bill of Rights, a multistakeholder process to specify how the principles in the Consumer Privacy Bill of Rights apply in particular business contexts, effective enforcement, and a commitment to increase interoperability with the privacy frameworks of our international partners.

- **A Consumer Privacy Bill of Rights**

  This document sets forth a Consumer Privacy Bill of Rights that, in the Administration's view, provides a baseline of clear protections for consumers and greater certainty for companies. The Administration will encourage stakeholders to implement the Consumer Privacy Bill of Rights through codes of conduct and will work with Congress to enact these rights through legislation. The Consumer Privacy Bill of Rights applies comprehensive, globally recognized Fair Information Practice Principles (FIPPs) to the interactive and highly interconnected environment in which we live and work today. Specifically, it provides for:

  - Individual Control: Consumers have a right to exercise control over what personal data companies collect from them and how they use it.

  - Transparency: Consumers have a right to easily understandable and accessible information about privacy and security practices.

  - Respect for Context: Consumers have a right to expect that companies will collect, use, and disclose personal data in ways that are consistent with the context in which consumers provide the data.

  - Security: Consumers have a right to secure and responsible handling of personal data.

  - Access and Accuracy: Consumers have a right to access and correct personal data in usable formats, in a manner that is appropriate to the sensitivity of the data and the risk of adverse consequences to consumers if the data is inaccurate.

  - Focused Collection: Consumers have a right to reasonable limits on the personal data that companies collect and retain.

  - Accountability: Consumers have a right to have personal data handled by companies with appropriate measures in place to assure they adhere to the Consumer Privacy Bill of Rights.

The Consumer Privacy Bill of Rights provides general principles that afford companies discretion in how they implement them. This flexibility will help promote innovation. Flexibility will also encourage effective privacy protections by allowing companies, informed by input from consumers and other stakeholders, to address the privacy issues that are likely to be most important to their customers and users, rather than requiring companies to adhere to a single, rigid set of requirements.

Enacting the Consumer Privacy Bill of Rights through Federal legislation would increase legal certainty for companies, strengthen consumer trust, and bolster the United States' ability to lead consumer data privacy engagements with our international partners. Even if Congress does not pass legislation, the Consumer Privacy Bill of Rights will serve as a template for privacy protections that increase consumer trust on the Internet and promote innovation.

- **Fostering Multistakeholder Processes to Develop Enforceable Codes of Conduct**

The Administration's framework outlines a multistakeholder process to produce enforceable codes of conduct that implement the Consumer Privacy Bill of Rights. The Administration will convene open, transparent forums in which stakeholders who share an interest in specific markets or business contexts will work toward consensus on appropriate, legally enforceable codes of conduct. Private sector participation will be voluntary and companies ultimately will choose whether to adopt a given code of conduct. The participation of a broad group of stakeholders, including consumer groups and privacy advocates, will help to ensure that codes of conduct lead to privacy solutions that consumers can easily use and understand. A single code of conduct for a given market or business context will provide consumers with more consistent privacy protections than is common today, when privacy practices and the information that consumers receive about them varies significantly from company to company.

- **Strengthening FTC Enforcement**

FTC enforcement is critical to ensuring that companies are accountable for adhering to their privacy commitments. Enforcement is also critical to ensuring that responsible companies are not disadvantaged by competitors who would play by different rules. As part of consumer data privacy legislation, the Administration encourages Congress to provide the FTC (and State Attorneys General) with specific authority to enforce the Consumer Privacy Bill of Rights.

- **Improving Global Interoperability**

The Administration's framework embraces the goal of increased international interoperability as a means to provide consistent, low-barrier rules for personal data in the user-driven and decentralized Internet environment. The two principles that underlie our approach to interoperability are mutual recognition and enforcement cooperation. Mutual recognition depends on effective enforcement and well-defined accountability mechanisms. Multistakeholder processes can provide scalable, flexible means of developing codes of conduct that simplify companies' compliance obligations. Enforcement cooperation helps to ensure that countries are able to protect their citizens' rights when personal data crosses national boundaries. These approaches will guide United States efforts to clarify data protections globally while ensuring the flexibility that is critical to innovation in the commercial world.

The Administration will implement this framework without delay. In the coming months, the Department of Commerce will work with other Federal agencies to convene stakeholders, including our international partners, to develop enforceable codes of conduct that build on the Consumer Privacy Bill of Rights.

# I. Introduction: Building on the Strength of the U.S. Consumer Data Privacy Framework

The Internet is integral to economic and social life in the United States and throughout the world. Networked technologies offer individuals nearly limitless ways to express themselves, form social connections, transact business, and organize politically. Networked technologies also spur innovation, enable new business models, and facilitate consumers' and companies' access to information, products, and services markets across the world.

An abundance of data, inexpensive processing power, and increasingly sophisticated analytical techniques drive innovation in our increasingly networked society. Political organizations and candidates for public office build powerful campaigns on data that individuals share about themselves and their political preferences. Data from social networks allows journalists and individuals to report and follow newsworthy events around the world as they unfold. Data plays a key role in the ability of government to stop identity thieves and protect public safety. Researchers use sets of medical data to identify public health issues and probe the causes of human diseases. Network operators use data from communications networks to identify events ranging from a severed fiber optic cable to power outages and the acts of malicious intruders. In addition, personal data fuels an advertising marketplace that brings many online services and sources of content to consumers for free.

Strengthening consumer data privacy protections in the United States is an important Administration priority.[1] Americans value privacy and expect protection from intrusions by both private and governmental actors. Strong privacy protections also are critical to sustaining the trust that nurtures Internet commerce and fuels innovation. Trust means the companies and technical systems on which we depend meet our expectations for privacy, security, and reliability.[2] In addition, United States leadership in consumer data privacy can help establish more flexible, innovation-enhancing privacy models among our international partners.[3]

---

1. This framework is concerned solely with how private-sector entities handle personal data in commercial settings. A separate set of constitutional and statutory protections apply to the government's access to data that is in the possession of private parties. In addition, the Privacy Act of 1974, Pub. L. No. 93-579 (5 U.S.C. § 552a), and implementing guidance from the Office of Management and Budget, *available at* http://www.whitehouse.gov/omb/privacy_general, govern the Federal government's handling of personally identifiable information. Both of these areas are beyond the scope of this document.

2. Throughout this document, "company" means any organization, corporation, trust, partnership, sole proprietorship, unincorporated association, or venture established to make a profit, or nonprofit entity, that collects, uses, discloses, stores, or transfers personal data in interstate commerce, to the extent such organizations are not subject to existing Federal data privacy laws.

3. *See, e.g.*, Remarks of Secretary of State Hillary Rodham Clinton, Release of Administration's International Strategy for Cyberspace (May 2011) ("Many of you representing the governments of other countries, as well as the private sector or foundations or civil society groups, share our commitment to ensuring that the Internet remains open, secure, free, not only for the 2 billion people who are now offline, but for the billions more who will be online in the years ahead.").

Preserving trust in the Internet economy protects and enhances substantial economic activity.[4] Online retail sales in the United States total $145 billion annually.[5] New uses of personal data in location services, protected by appropriate privacy and security safeguards, could create important business opportunities.[6] Moreover, the United States is a world leader in exporting cloud computing, location-based services, and other innovative services. To preserve these economic benefits, consumers must continue to trust networked technologies. Strengthening consumer data privacy protections will help to achieve this goal.

Preserving trust also is necessary to realize the full social and cultural benefits of networked technologies. When companies use personal data in ways that are inconsistent with the circumstances under which consumers disclosed the data, however, they may undermine trust. For example, individuals who actively share information with their friends, family, colleagues, and the general public through websites and online social networking sites may not be aware of the ways those services, third parties, and their own associates may use information about them. Unauthorized disclosure of sensitive information can violate individual rights, cause injury or discrimination based on sensitive personal attributes, lead to actions and decisions taken in response to misleading or inaccurate information, and contribute to costly and potentially life-disrupting identity theft.[7] Protecting Americans' privacy by preventing identity theft and prosecuting identity thieves is an important focus for the Administration.

The existing consumer data privacy framework in the United States is flexible and effectively addresses some consumer data privacy challenges in the digital age. This framework consists of industry best practices, FTC enforcement, and a network of chief privacy officers and other privacy professionals who develop privacy practices that adapt to changes in technology and business models and create a growing culture of privacy awareness within companies. Much of the personal data used on the Internet, however, is not subject to comprehensive Federal statutory protection, because most Federal data privacy statutes apply only to specific sectors, such as healthcare, education, communications, and financial services or, in the case of online data collection, to children. The Administration believes that filling gaps in the existing framework will promote more consistent responses to privacy concerns across the wide range of environments in which individuals have access to networked technologies and in which a broad array of companies collect and use personal data. The Administration, however, does not recommend modifying the existing Federal statutes that apply to specific sectors unless they set inconsistent standards for related technologies. Instead, the Administration supports legislation that would supplement the existing framework and extend baseline protections to the sectors that existing Federal statutes do not cover.

---

4. President Barack Obama, *International Strategy for Cyberspace*, at 8, May 2011, http://www.whitehouse.gov/sites/default/files/rss_viewer/international_strategy_for_cyberspace.pdf.

5. U.S. Census Bureau, *E-Stats*, May 26, 2011, http://www.census.gov/econ/estats/2009/2009reportfinal.pdf, at 1.

6. McKinsey Global Institute, *Big Data: The Next Frontier for Innovation, Competition, and Productivity*, at 94-95, May 2011, http://www.mckinsey.com/mgi/publications/big_data/pdfs/MGI_big_data_full_report.pdf. The National Institute of Standards and Technology (NIST) has identified five essential characteristics of cloud computing: on-demand self-service, broad network access, resource pooling, rapid elasticity, and measured service. Peter Mell and Tim Gance, The NIST Definition of Cloud Computing, version 15, Oct. 7, 2009, http://csrc.nist.gov/groups/SNS/cloud-computing/cloud-def-v15.doc.

7. Recently, identity theft alone was estimated to cause economic losses of more than $15 billion in a single year. Fed. Trade Comm'n, 2006 Identity Theft Survey Report (2007), *available at* http://www.ftc.gov/os/2007/11/SynovateFinalReportIDTheft2006.pdf.

The comprehensive consumer data privacy framework set forth here will provide clearer protections for consumers. It will also provide greater certainty for companies while promoting innovation and minimizing compliance costs (consistent with the goals of Executive Order 13563, "Improving Regulation and Regulatory Review"). The framework provides consumers who want to understand and control how personal data flows in the digital economy with better tools to do so. The proposal ensures that companies striving to meet consumers' expectations have more effective ways of engaging consumers and policymakers. This will help companies to determine which personal data practices consumers find unobjectionable and which ones they find invasive. Finally, the Administration's consumer data privacy framework improves our global competitiveness by promoting international policy frameworks that reflect how consumers and companies actually use networked technologies.

As a world leader in Internet innovation, the United States has both the responsibility and incentive to help establish forward-looking privacy policy models that foster innovation and preserve basic privacy rights. The Administration's framework for consumer data privacy offers a path toward achieving these goals. It is based on the following key elements:

- A **Consumer Privacy Bill of Rights**, setting forth individual rights and corresponding obligations of companies in connection with personal data. These consumer rights are based on U.S.-developed and globally recognized Fair Information Practice Principles (FIPPs), articulated in terms that apply to the dynamic environment of the Internet age;

- **Enforceable codes of conduct**, developed through **multistakeholder processes**, to form the basis for specifying what the Consumer Privacy Bill of Rights requires in particular business contexts;

- Federal Trade Commission (FTC) **enforcement** of consumers' data privacy rights through its authority to prohibit unfair or deceptive acts or practices; and

- Increasing **global interoperability** between the U.S. consumer data privacy framework and other countries' frameworks, through mutual recognition, the development of codes of conduct through multistakeholder processes, and enforcement cooperation can reduce barriers to the flow of information.

*Consumer Data Privacy in a Networked World* builds on the recommendations of the Department of Commerce Internet Policy Task Force's December 2010 report, *Commercial Data Privacy and Innovation in the Internet Economy: A Dynamic Policy Framework* ("Privacy and Innovation Green Paper").[8] The Internet Policy Task Force developed the recommendations in the Privacy and Innovation Green Paper by engaging with stakeholders—companies, trade groups, privacy advocates, academics, State Attorneys General, Federal civil and criminal law enforcement representatives, and international partners—through a public symposium, written comments, public speeches and presentations, and informal meetings. More than 100 stakeholders subsequently submitted written comments on the Privacy and Innovation Green Paper. These comments provided the Administration with invaluable feedback during the development of *Consumer Data Privacy in a Networked World*. The Administration gratefully acknowledges the time and resources stakeholders devoted to this issue. Their ongoing engagement will be critical to implementing the framework successfully.

---

8. Department of Commerce, *Commercial Data Privacy and Innovation in the Internet Economy: Dynamic Policy Framework*, Dec. 2010, *available* at http://www.ntia.doc.gov/report/2010/commercial-data-privacy-and-innovation-internet-economy-dynamic-policy-framework.

# II. Defining a Consumer Privacy Bill of Rights

Strengthening consumer data privacy protections and promoting innovation require privacy protections that are comprehensive, actionable, and flexible. The United States pioneered the FIPPs in the 1970s, and they have become the globally recognized foundations for privacy protection. The United States has embraced FIPPs by incorporating them into sector-specific privacy laws and applying them to personal data that Federal agencies collect. FIPPs also are a foundation for numerous international data privacy frameworks.[9] These principles continue to provide a solid foundation for consumer data privacy protection, despite far-reaching changes in companies' ability to collect, store, and analyze personal data.

The Consumer Privacy Bill of Rights applies FIPPs to an environment in which processing of data about individuals is far more decentralized and pervasive than it was when FIPPs were initially developed. Large corporations and government agencies collecting information for relatively static databases are no longer typical of personal data collectors and processors. The world is far more varied and dynamic. Companies process increasing quantities of personal data for a widening array of purposes. Consumers increasingly exchange personal data in active ways through channels such as online social networks and personal blogs. The reuse of personal data can be an important source of innovation that brings benefits to consumers but also raises difficult questions about privacy. The central challenge in this environment is to protect consumers' privacy expectations while providing companies with the certainty they need to continue to innovate.[10]

To meet this challenge, the Consumer Privacy Bill of Rights carries FIPPs forward in two ways. First, it affirms a set of consumer rights that inform consumers of what they should expect of companies that handle personal data. The Consumer Privacy Bill of Rights also recognizes that consumers have certain responsibilities to protect their privacy as they engage in an increasingly networked society. Second, the Consumer Privacy Bill of Rights reflects the FIPPs in a way that emphasizes the importance of context in their application.[11] Key elements of context include the goals or purposes that consumers can expect

---

9. As noted in the Privacy and Innovation Green Paper (p. 11):

> In 1973, the Department of Health, Education, and Welfare (HEW) released its report, *Records, Computers, and the Rights of Citizens*, which outlined a Code of Fair Information Practices that would create "safeguard requirements" for certain "automated personal data systems" maintained by the Federal Government. This Code of Fair Information Practices, now commonly referred to as fair information practice principles (FIPPs), established the framework on which much privacy policy would be built.

Examples of FIPPs-based international frameworks include the Organisation for Economic Co-operation and Development *Guidelines on the Protection of Privacy and Transborder Flows of Personal Data* and the Asia-Pacific Economic Cooperation *Privacy Framework*. The Privacy and Innovation Green Paper proposed for consideration the following set of FIPPs: transparency, individual participation, purpose specification, data minimization, use limitation, data quality and integrity, security, and accountability and auditing.

10. As the Privacy and Innovation Green Paper noted, "New devices and applications allow the collection and use of personal information in ways that, at times, can be contrary to many consumers' privacy expectations." Department of Commerce, Privacy and Innovation Green Paper, at i (statement of Commerce Secretary Gary Locke).

11. For a comparison of the Consumer Privacy Bill of Rights to other statements of the FIPPs, see Appendix B.

to achieve by using a company's products or services, the services that the companies actually provide, the personal data exchanges that are necessary to provide these services, and whether a company's customers include children and adolescents. Context should shape the balance and relative emphasis of particular principles in the Consumer Privacy Bill of Rights.

The Consumer Privacy Bill of Rights advances these objectives by holding that consumers have a right to:

- Individual Control
- Transparency
- Respect for Context
- Security
- Access and Accuracy
- Focused Collection
- Accountability

The Consumer Privacy Bill of Rights applies to commercial uses of personal data. This term refers to any data, including aggregations of data, which is linkable to a specific individual.[12] Personal data may include data that is linked to a specific computer or other device. For example, an identifier on a smartphone or family computer that is used to build a usage profile is personal data. This definition provides the flexibility that is necessary to capture the many kinds of data about consumers that commercial entities collect, use, and disclose.

The remainder of this section provides the full statement of the Consumer Privacy Bill of Rights and explains the rationale for the rights and obligations under each principle.

---

12. This definition is similar to the Federal Government's definition of "personally identifiable information":

> [I]nformation that can be used to distinguish or trace an individual's identity, either alone or when combined with other personal or identifying information that is linked or linkable to a specific individual. The definition of PII is not anchored to any single category of information or technology. Rather, it requires a case-by-case assessment of the specific risk that an individual can be identified.

Peter R. Orszag, Memorandum for the Heads of Executive Departments and Agencies, Guidance for Agency Use of Third-Party Websites and Applications, at 8 (Appendix), June 25, 2010, http://www.whitehouse.gov/sites/default/files/omb/assets/memoranda_2010/m10-23.pdf.

> 1. **Individual Control: Consumers have a right to exercise control over what personal data companies collect from them and how they use it.** Companies should provide consumers appropriate control over the personal data that consumers share with others and over how companies collect, use, or disclose personal data. Companies should enable these choices by providing consumers with easily used and accessible mechanisms that reflect the scale, scope, and sensitivity of the personal data that they collect, use, or disclose, as well as the sensitivity of the uses they make of personal data. Companies should offer consumers clear and simple choices, presented at times and in ways that enable consumers to make meaningful decisions about personal data collection, use, and disclosure. Companies should offer consumers means to withdraw or limit consent that are as accessible and easily used as the methods for granting consent in the first place.

The Individual Control principle has two dimensions. First, at the time of collection, companies should present choices about data sharing, collection, use, and disclosure that are appropriate for the scale, scope, and sensitivity of personal data in question. For example, companies that have access to significant portions of individuals' Internet usage histories, such as search engines, ad networks, and online social networks, can build detailed profiles of individual behavior over time. These profiles may be broad in scope and large in scale, and they may contain sensitive information, such as personal health or financial data.[13] In these cases, choice mechanisms that are simple and prominent and offer fine-grained control of personal data use and disclosure may be appropriate. By contrast, services that do not collect information that is reasonably linkable to individuals may offer accordingly limited choices.

In any event, a company that deals directly with consumers should give them appropriate choices about what personal data the company collects, irrespective of whether the company uses the data itself or discloses it to third parties. When consumer-facing companies contract with third parties that gather personal data directly from consumers (as is the case with much online advertising), they should be diligent in inquiring about how those third parties use personal data and whether they provide consumers with appropriate choices about collection, use, and disclosure. The Administration also encourages consumer-facing companies to act as stewards of personal data that they and their business partners collect from consumers. Consumer-facing companies should seek ways to recognize consumer choices through mechanisms that are simple, persistent, and scalable from the consumer's perspective.

Third parties should also offer choices about personal data collection that are appropriate for the scale, scope, and sensitivity of data they collect. The focal point for much of the debate about third-party personal data collection in recent years is online behavioral advertising—the practice of collecting

---

13. "Scope" refers to the range of activities or interests as well as the time period that is reflected in a dataset. "Scale" refers to the number of individuals whose activities are in a dataset.

information about consumers' online interests in order to deliver targeted advertising to them.[14] This system of advertising revolves around ad networks that can track individual consumers—or at least their devices—across different websites. When organized according to unique identifiers, this data can provide a potentially wide-ranging view of individual use of the Internet. These individual behavioral profiles allow advertisers to target ads based on inferences about individual interests, as revealed by Internet use. Targeted ads are generally more valuable and efficient than purely contextual ads and provide revenue that supports an array of free online content and services.[15] However, many consumers and privacy advocates find tracking and the advertising practices that it enables invade their expectations of privacy.[16]

The Administration recognizes that the ultimate uses of personal data that third parties, such as ad networks, collect affect the privacy interests at stake. As a result, these uses of personal data should help to shape the range of appropriate individual control options. For example, a company that uses personal data only to calculate statistics about how consumers use its services may not implicate significant consumer privacy interests and may not need to provide consumers with ways to prevent data collection for this purpose. Even if the company collects and stores some personal data for some uses, it may not need to provide consumers with a sophisticated array of choices about collection. In the case of online advertising, for instance, verifying ad delivery and preventing a consumer from seeing the same ad many times over may require some personal data collection. But personal data collected only for these statistical purposes may not require the assembly of extensive, long-lived individual profiles and may not require extensive options for control.

Innovative technology can help to expand the range of user control. It is increasingly common for Internet companies that have direct relationships with consumers to offer detailed privacy settings that allow individuals to exercise greater control over what personal data the companies collect, and when. In addition, privacy-enhancing technologies such as the "Do Not Track" mechanism allow consumers to exercise some control over how third parties use personal data or whether they receive it at all. For example, prompted by the FTC,[17] members of the online advertising industry developed self-regulatory principles based on the FIPPs, a common interface to alert consumers of the presence of third party ads and to direct them to more information about the relevant ad network, and a common mechanism to

---

14.  *See* FTC, *Self-Regulatory Principles for Online Behavioral Advertising* (staff report), at 2, Feb. 2009 (stating that online behavioral advertising "involves the tracking of consumers' online activities in order to deliver tailored advertising").

15.  According to one study, behaviorally targeted ads are worth significantly more than non-targeted ads. *See* Howard Beales, *The Value of Behavioral Targeting*, at 3, Mar. 24, 2010 (finding, based on data provided by ad networks, that behaviorally targeted ad rates in 2009 were 2.68 times greater than non-targeted ad rates), http://www.networkadvertising.org/pdfs/Beales_NAI_Study.pdf; FTC, *Protecting Consumer Privacy in an Era of Rapid Change: A Proposed Framework for Businesses and Policymakers* (preliminary staff report), at 24, Dec. 2010 (reporting that FTC privacy roundtable participants discussed that "the more information that is known about a consumer, the more a company will pay to deliver a precisely-targeted advertisement to him") ("FTC Staff Report").

16.  See Aleecia M. McDonald and Lorrie Faith Cranor, *Americans' Attitudes About Internet Behavioral Advertising Practices*, Proceedings of the 9th Annual ACM Workshop on Privacy in the Electronic Society (WPES) (2010).

17.  *See generally* FTC, *Self-Regulatory Principles for Online Behavioral Advertising* (staff report), Feb. 2009.

allow consumers to opt out of targeted advertising by individual ad networks.[18] A variety of other actors, including browser vendors, software developers, and standards-setting organizations, are developing "Do Not Track" mechanisms that allow consumers to exercise some control over whether third parties receive personal data. All of these mechanisms show promise. However, they require further develop-ment to ensure they are easy to use, strike a balance with innovative uses of personal data, take public safety interests into account, and present consumers with a clear picture of the potential costs and benefits of limiting personal data collection.

As third parties become further removed from direct interactions with consumers, it may be more difficult for them to provide consumers with meaningful control over data collection. Data brokers, for example, aggregate personal data from multiple sources, often without interacting with consumers at all. Such companies face a challenge in providing effective mechanisms for individual control because consumers might not know that these third parties exist. Moreover, some data brokers collect court records, news reports, property records, and other data that is in the public record. The rights of free-dom of speech and freedom of the press involved in the collection and use of these documents must be balanced with the need for transparency to individuals about how data about them is collected, used, and disseminated and the opportunity for individuals to access and correct data that has been collected about them.

Still, data brokers and other companies that collect personal data without direct consumer interactions or a reasonably detectable presence in consumer-facing activities should seek innovative ways to provide consumers with effective Individual Control. If it is impractical to provide Individual Control, these com-panies should ensure that they implement other elements of the Consumer Privacy Bill of Rights in ways that adequately protect consumers' privacy. For example, to provide sufficient privacy protections, such companies may need to go to extra lengths to implement other principles such as Transparency—by providing clear, public explanations of the roles they play in commercial uses of personal data—as well as providing appropriate use controls once information is collected under the Access and Accuracy and Accountability principles to compensate for the lack of a direct consumer relationship.

The second dimension of Individual Control is consumer responsibility. In a growing number of cases, such as online social networks, the use of personal data begins with individuals' decisions to choose privacy settings and to share personal data with others. In such contexts, consumers should evaluate their choices and take responsibility for the ones that they make. Control over the initial act of sharing is critical. Consumers should take responsibility for those decisions, just as companies that participate in and benefit from this sharing should provide usable tools and clear explanations to enable consumers to make meaningful choices.

The Individual Control principle also recognizes that consumers' privacy interests in personal data persist throughout their relationships with a company. Accordingly, this principle includes a right to withdraw consent to use personal data that the company controls. Companies should provide means of with-

---

18. *See* AboutAds.info, *Self-Regulatory Principles for Online Behavioral Advertising,* http://www.aboutads.info/resource/download/seven-principles-07-01-09.pdf (July 2009); Interactive Advertising Bureau, Comment on the Privacy and Innovation Green Paper (Attachment B) (explaining online advertisers' system for directing users to ad networks' privacy policies and opt-outs).

drawing consent that are on equal footing with ways they obtain consent. For example, if consumers grant consent through a single action on their computers, they should be able to withdraw consent in a similar fashion.[19]

There are three practical limits to the right to withdraw consent. First, it presumes that consumers have an ongoing relationship with a company. This relationship could be minimal, such as a consumer establishing an account for a single transaction; or it may be as extensive as many financial transactions spanning many years. Nonetheless, the company must have a way to effect a withdrawal of consent to the extent the company has associated and retained data with an individual. Conversely, data that a company cannot reasonably associate with an individual is not subject to the right to withdraw consent. Second, the obligation to respect a consumer's withdrawal of consent only extends to data that the company has under its control. Third, the Individual Control principle does not call for companies to permit withdrawal of consent for personal data that they collected before implementing the Consumer Privacy Bill of Rights, unless they made such a commitment at the time of collection.

> 2. **TRANSPARENCY: Consumers have a right to easily understandable and accessible information about privacy and security practices.** At times and in places that are most useful to enabling consumers to gain a meaningful understanding of privacy risks and the ability to exercise Individual Control, companies should provide clear descriptions of what personal data they collect, why they need the data, how they will use it, when they will delete the data or de-identify it from consumers, and whether and for what purposes they may share personal data with third parties.

Plain language statements about personal data collection, use, disclosure, and retention help consumers understand the terms surrounding commercial interactions. Companies should make these statements visible to consumers when they are most relevant to understanding privacy risks and easily accessible when called for.

Personal data uses that are not consistent with the context of a company-to-consumer transaction or relationship deserve more prominent disclosure than uses that are integral to or commonly accepted in that context. Privacy notices that distinguish personal data uses along these lines will better inform consumers of personal data uses that they have not anticipated, compared to many current privacy notices that generally give equal emphasis to all potential personal data uses.[20] Such notices will give privacy-conscious consumers easy access to information that is relevant to them. They may also promote greater consistency in disclosures by companies in a given market and attract the attention of consumers who ordinarily would ignore privacy notices, potentially making privacy practices a more salient point of competition among different products and services.

---

19. The obligation to provide these choices should be read in conjunction with the Access and Accuracy principle discussed below.

20. *See* Assistant Secretary for Communications and Information Lawrence E. Strickling, Testimony Before the Senate Committee on Commerce, Science, and Transportation, Mar. 16, 2011, at 2-3.

In addition, companies should provide notice in a form that is easy to read on the devices that consumers actually use to access their services. In particular, mobile devices have small screens that make reading full privacy notices effectively impossible. Companies should therefore strive to present mobile consumers with the most relevant information in a manner that takes into account mobile device characteristics, such as small display sizes and privacy risks that are specific to mobile devices.

Finally, companies that do not interact directly with consumers—such as the data brokers discussed above—need to make available explicit explanations of how they acquire, use, and disclose personal data. These companies may need to compensate for the lack of a direct relationship when making these explanations available, for example by posting them on their websites or other publicly accessible locations. Moreover, companies that have first-party relationships with consumers should disclose specifically the purpose(s) for which they provide personal data to third parties, help consumers to understand the nature of those third parties' activities, and whether those third parties are bound to limit their use of the data to achieving those purposes. This gives consumers a more tractable task of assessing whether to engage with a single entity, rather than trying to understand what personal data third parties—potentially dozens, or even hundreds—receive and how they use it. Similarly, first parties could create greater transparency by disclosing what kinds of personal data they obtain from third parties, who the third parties are, and how they use this data. This level of transparency may also facilitate the development within the private sector of innovative privacy-enhancing technologies and guidance that consumers can use to protect their privacy.

> 3. **RESPECT FOR CONTEXT: Consumers have a right to expect that companies will collect, use, and disclose personal data in ways that are consistent with the context in which consumers provide the data.** Companies should limit their use and disclosure of personal data to those purposes that are consistent with both the relationship that they have with consumers and the context in which consumers originally disclosed the data, unless required by law to do otherwise. If companies will use or disclose personal data for other purposes, they should provide heightened Transparency and Individual Choice by disclosing these other purposes in a manner that is prominent and easily actionable by consumers at the time of data collection. If, subsequent to collection, companies decide to use or disclose personal data for purposes that are inconsistent with the context in which the data was disclosed, they must provide heightened measures of Transparency and Individual Choice. Finally, the age and familiarity with technology of consumers who engage with a company are important elements of context. Companies should fulfill the obligations under this principle in ways that are appropriate for the age and sophistication of consumers. In particular, the principles in the Consumer Privacy Bill of Rights may require greater protections for personal data obtained from children and teenagers than for adults.

Respect for Context distinguishes personal data uses on the basis of how closely they relate to the purposes for which consumers use a service or application as well as the business processes necessary to provide the service or application.[21] The Respect for Context principle calls on companies that collect data to act as stewards of data in ways that respect their consumers. This principle derives from two principles commonly found in statements of the FIPPs. The first principle, purpose specification, states that companies should specify at the time of collection the purposes for which they collect personal data. Second, the use limitation principle holds that companies should use personal data only to fulfill those specific purposes.

The Respect for Context principle adapts these well-established principles in two ways. First, Respect for Context provides a substantive standard to guide companies' decisions about their basic personal data practices. Generally speaking, companies should limit personal data uses to fulfilling purposes that are consistent with the context in which consumers disclose personal data. Second, while this principle emphasizes the importance of the relationship between a consumer and a company at the time consumers disclose data, it also recognizes that this relationship may change over time in ways not foreseeable at the time of collection. Such adaptive uses of personal data may be the source of innovations that benefit consumers. However, companies must provide appropriate levels of transparency and individual choice—which may be more stringent than was necessary at the time of collection—before reusing personal data.

Applying the Consumer Privacy Bill of Rights in a context-specific manner provides companies flexibility but also requires them to consider carefully what consumers are likely to understand about their data practices based on the products and services they offer, how the companies themselves explain the roles of personal data in delivering them, research on consumers' attitudes and understandings, and feedback from consumers. Context should help to determine which personal data uses are likely to raise the greatest consumer privacy concerns. The company-to-consumer relationship should guide companies' decisions about which uses of personal data they will make most prominent in privacy notices. For

---

21.  Several commenters on the Privacy and Innovation Green Paper emphasized the importance of context in applying FIPPs. *See, e.g.,* AT&T Comment on the Privacy and Innovation Green Paper, at 7, Jan. 28, 2011 ("FIPPs are usefully expressed as generalized policy guides that should shape the multi-stakeholder collaborative processes to develop flexible and contextualized codes of practice for particular industries."); Centre for Information Policy Leadership Comment on the Privacy and Innovation Green Paper, at 3, Jan. 28, 2011 ("Principles of fair information practices should be applied within a contextual framework, and not in a rigid or fixed way."); Google Comment on the Privacy and Innovation Green Paper, at 6, Jan. 28, 2011 ("In particular, FIPPs must be flexible enough to take account of the spectrum of identifiability, linkability, and sensitivity of various data in various contexts."); Intel Comment on the Privacy and Innovation Green Paper, at 4 ("[M]any of the issues present in a privacy regulatory scheme are highly contextual."); Intuit Comment on the Privacy and Innovation Green Paper, at 9 ("It is the use of the information as well as its characteristics that should inform our treatment of it. Context is crucial."); Helen Nissenbaum, Kenneth Farrall, and Finn Brunton, Comment on the Privacy and Innovation Green Paper, at 2-3 (recommending consideration of context as a source of "baseline substantive constraints on data practices following the model of current US sectoral privacy regulation"); Online Publishers Association Comment on the Privacy and Innovation Green Paper, at 6 ("Online publishers share a direct and trusted relationship with visitors to their sites. In the context of this relationship, OPA members sometimes collect and use information to target and deliver the online advertising that subsidizes production of quality digital content."); TRUSTe Comment on the Privacy and Innovation Green Paper, at 2 ("We view privacy as inherently contextual; disclosure obligations will differ depending on the context of the interaction."). Current scholarship also emphasizes the importance of the relationship between context and privacy. *See* Helen Nissenbaum, *Privacy in Context: Technology, Policy, and the Integrity of Social Life* (2009).

example, online retailers need to disclose consumers' names and home addresses to shippers in order to fulfill customers' orders. This disclosure is obvious from the context of the consumer-retailer relationship. Retailers do not need to provide prominent notice of the practice (though they should disclose it in their full privacy notices); companies may infer that consumers have agreed to the disclosure based on the consumers' actions in placing the order and a widespread understanding of the product delivery process.

Several categories of data practices are both common to many contexts and integral to companies' operations. The example above falls into the more general category of product and service fulfillment; companies may infer consent to use and disclose personal data to achieve objectives that consumers have specifically requested, as long as there is a common understanding of the service. Similarly, companies may infer consent to use personal data to conduct marketing in the context of most first-party relationships, given the familiarity of this activity in digital and in-person commerce, the visibility of this kind of marketing, the presence of an easily identifiable party to contact to provide feedback, and consumers' opportunity to end their relationship with a company if they are dissatisfied with it. In addition, companies collect and use personal data for purposes that are common, even if they may not be well known to consumers. For example, analyzing how consumers use a service in order to improve it, preventing fraud, complying with law enforcement orders and other legal obligations, and protecting intellectual property all have been basic elements of doing business and meeting companies' legal obligations.[22] Companies should be able to infer consumer consent to collect personal data for these limited purposes, consistent with the other principles in the Consumer Privacy Bill of Rights.

In other cases, context should guide decisions about which opportunities for consumer control are reasonable for companies to provide and also meaningful to consumers. Information and choices that are meaningful to consumers in one context may be largely irrelevant in others. For example, consider a hypothetical game application for a mobile device that allows consumers to save the game's state, so that they can resume playing after a break. The hypothetical company that provides this game collects the unique identifier of each user's mobile device in order to provide this "save" function. Collecting the mobile device's unique identifier for this purpose may be consistent with the "save" function and consumers' decisions to use it, particularly if the company uses identifiers only for this purpose. If the company provides consumers' unique device identifiers to third parties for purposes such as online behavioral advertising, however, the company should notify consumers and allow them to prevent the disclosure of personal data.

The sophistication of a company's consumers is also a critical element of context. In particular, the privacy framework may require a different degree of protection for children's and teenagers' privacy interests from the protections afforded to adults due to the unique characteristics of these age groups. Children may be particularly susceptible to privacy harms. Currently, the Children's Online Privacy Protection Act (COPPA) and the FTC's implementing regulations provide strong protections by requiring online

---

22. This list of practices that are common to many contexts is similar to the "commonly accepted practices" that FTC staff identified in its 2010 report. *See* FTC Staff Report at 53-54. In the Administration's view, protecting intellectual property is so widespread and necessary to many companies that they should be able to infer consent to achieve this objective. Several commenters on the Department of Commerce's Privacy and Information Green Paper encouraged the Administration to recognize such practices in order to provide certainty for companies and to give greater prominence to choices that consumers are more likely to find meaningful.

services that are directed to children, or that know that they are collecting personal data from children, to obtain verifiable parental consent before they collect such data.[23] Online services that are "directed to" children must meet this same standard. The Administration looks forward to exploring with stakeholders whether more stringent applications of the Consumer Privacy Bill of Rights—such as an agreement not to create individual profiles about children, even if online services obtain the necessary consent to collect personal data—are appropriate to protect children's privacy.

The terms governing a company-to-consumer relationship are another key element of context. In particular, advertising supports innovative new services and helps to provide consumers with free access to a broad array of online services and applications. The Respect for Context principle does not foreclose any particular ad-based business models. Rather, the Respect for Context principle requires companies to recognize that different business models based on different personal data raise different privacy risks. A company should clearly inform consumers of what they are getting in exchange for the personal data they provide. The Administration also encourages companies engaged in online advertising to refrain from collecting, using, or disclosing personal data that may be used to make decisions regarding employment, credit, and insurance eligibility or similar matters that may have significant adverse consequences to consumers. Collecting data for such sensitive uses is at odds with the contextually well-defined purposes of generating revenue and providing consumers with ads that they are more likely to find relevant. Such practices also may be at odds with the norm of responsible data stewardship that the Respect for Context principle encourages.

Consider, for example, an online social networking service whose users disclose biographical information when creating an account and provide information about their social contacts and interests by including friends, business associates, and companies in their networks. As consumers use the service, they may generate large amounts of information that is associated with their identity on the online social network, including written updates, photos, videos, and location information. Consumers make affirmative choices to share this information with members of their online social networks. These disclosures are all integral to the company providing its social networking service. Furthermore, it is reasonable for the company to reveal at least some of these details to other members in order to help them form new connections.

Whether the online social networking service provider will use this information, and for what purposes, may be less clear from the context that consumers experience. The personal data that consumers generate may be valuable for improving the service, selling online advertising, or assembling individual profiles that the company provides to third parties. These uses fall along a continuum that starts at the core context of consumers engaging online with a group of associates. Consumers expect the company to improve its services. The company does not need to seek affirmative consent each time it uses existing data to improve a service, or even creates a new service, provided that these new uses of personal data are consistent with what users come to expect in a social networking context.

Suppose that the company leases individual profile information to third parties, such as information brokers. Respect for Context may not require the company to specify each use that a recipient might

---

23. *See* Children's Online Privacy Protection Act, Pub. L. 105-277 (codified at 15 U.S.C. §§ 6501-6506) *and* FTC, Children's Online Protection Rule, 16 C.F.R. Part 312. COPPA defines "child" to mean "an individual under the age of 13." 15 U.S.C. § 6501(1).

make of this data, but, at a minimum, it may require the company to state prominently and explicitly that it discloses personal data to third parties who may further aggregate and use this data for other purposes. The Respect for Context principle, in combination with other principles in the Consumer Privacy Bill of Rights, also calls on the company to provide consumers with meaningful opportunities to prevent these disclosures.

> **4. SECURITY: Consumers have a right to secure and responsible handling of personal data.** Companies should assess the privacy and security risks associated with their personal data practices and maintain reasonable safeguards to control risks such as loss; unauthorized access, use, destruction, or modification; and improper disclosure.

Technologies and procedures that keep personal data secure are essential to protecting consumer privacy. Security failures involving personal data, whether resulting from accidents or deliberate attacks, can cause harms that range from embarrassment to financial loss and physical harm. Companies that lose control of personal data may suffer reputational harm as well as financial losses if business partners or consumers end their relationships after a security breach. These consequences provide companies with significant incentives to keep personal data secure. The security precautions that are appropriate for a given company will depend on its lines of business, the kinds of personal data it collects, the likelihood of harm to consumers, and many other factors.

The Security principle recognizes these needs. It gives companies the discretion to choose technologies and procedures that best fit the scale and scope of the personal data that they maintain, subject to their obligations under any applicable data security statutes, including their duties to notify consumers and law enforcement agencies if the security of data about them is breached, and their commitments to adopt reasonable security practices.

> **5. ACCESS AND ACCURACY: Consumers have a right to access and correct personal data in usable formats, in a manner that is appropriate to the sensitivity of the data and the risk of adverse consequences to consumers if the data is inaccurate.** Companies should use reasonable measures to ensure they maintain accurate personal data. Companies also should provide consumers with reasonable access to personal data that they collect or maintain about them, as well as the appropriate means and opportunity to correct inaccurate data or request its deletion or use limitation. Companies that handle personal data should construe this principle in a manner consistent with freedom of expression and freedom of the press. In determining what measures they may use to maintain accuracy and to provide access, correction, deletion, or suppression capabilities to consumers, companies may also consider the scale, scope, and sensitivity of the personal data that they collect or maintain and the likelihood that its use may expose consumers to financial, physical, or other material harm.

An increasingly diverse array of entities uses personal data to make decisions that affect consumers in ways ranging from the ads they see online to their candidacy for employment. Outside of sectors covered by specific Federal privacy laws, such as the Health Insurance Portability and Accountability Act (HIPAA) and the Fair Credit Reporting Act, consumers do not currently have the right to access and correct this data. The Administration is committed to publishing data on the Internet in machine-readable formats to advance the goals of innovation, transparency, participation, and collaboration. For example, to promote innovation and efficiency in the delivery of electricity, the Administration supports providing consumers with timely access to energy usage data in standardized, machine-readable formats over the Internet.[24] Similarly, the expanded use of health IT, including patients' access to health data through electronic health records, is a key element of the Administration's innovation strategy.[25] Comprehensive privacy and security safeguards, tailored for both contexts, are fundamental to both strategies.

Providing consumers with access to information about them in usable formats holds similar promise in the commercial arena. To help consumers make more informed choices, the Administration encourages companies to make personal data available in useful formats to the properly authenticated individuals over the Internet.[26]

The Access and Accuracy principle recognizes that the use of inaccurate personal data may lead to a range of harms. The risk of these harms, in addition to the scale, scope, and sensitivity of personal data that a company retains, help to determine what kinds of access and correction facilities may be reasonable in a given context. As a result, this principle does not distinguish between companies that are consumer-facing and those that are not. In all cases, however, the mechanisms that companies use to provide consumers with access to data about them should not create additional privacy or security risks.

United States Constitutional law has long recognized that privacy interests co-exist alongside fundamental First Amendment rights to freedom of speech, freedom of the press, and freedom of association. Individuals and members of the press exercising their free speech rights may well speak about other individuals and include personal information in their speech. The Access and Accuracy principle should therefore be interpreted with full respect for First Amendment values, especially for non-commercial speakers and individuals exercising freedom of the press.

---

24.  National Science and Technology Council, *A Policy Framework for the 21st Century Grid: Enabling Our Secure Energy Future*, at 41, 46, June 2011, *available* at http://www.whitehouse.gov/sites/default/files/microsites/ostp/nstc-smart-grid-june2011.pdf.

25.  See The White House, *A Strategy for American Innovation: A Strategy for American Innovation: Securing Our Economic Growth and Prosperity*, Feb. 2011, http://www.whitehouse.gov/innovation/strategy; Department of Health and Human Services, Final Rule on Electronic Health Record Incentive Program, 75 Fed. Reg. 44314, July 28, 2010.

26.  *See* Memorandum for the Heads of Executive Departments and Agencies, "Informing Consumers Through Smart Disclosure," *available at* http://www.whitehouse.gov/sites/default/files/omb/inforeg/for-agencies/informing-consumers-through-smart-disclosure.pdf ("To the extent practicable and subject to valid restrictions, agencies should publish information online in an open format that can be retrieved, downloaded, indexed, and searched by commonly used Web search applications. An open format is one that is platform independent, machine readable, and made available to the public without restriction that would impede the re-use of that information."); M-10-06, Memorandum for the Heads of Executive Departments and Agencies, "Open Government Directive," *available at* http://www.whitehouse.gov/sites/default/files/omb/assets/memoranda_2010/m10-06.pdf ("Machine readable data are digital information stored in a format enabling the information to be processed and analyzed by computer. These formats allow electronic data to be as usable as possible.").

6.  **FOCUSED COLLECTION: Consumers have a right to reasonable limits on the personal data that companies collect and retain.** Companies should collect only as much personal data as they need to accomplish purposes specified under the Respect for Context principle. Companies should securely dispose of or de-identify personal data once they no longer need it, unless they are under a legal obligation to do otherwise.

The Focused Collection principle holds that companies should engage in considered decisions about the kinds of data they need to collect to accomplish specific purposes. For example, the hypothetical game company referenced above that collects the unique identifier of each user's mobile device in order to provide a "save" function should consider whether it must use the mobile device identifier or whether a less broadly linkable identifier would work as well. Nevertheless, as discussed under the Respect for Context principle, companies may find new uses for personal data after they collect it, provided they take appropriate measures of transparency and individual choice. The Focused Collection principle does not relieve companies of any independent legal obligations, including law enforcement orders, that require them to retain personal data.

Wide-ranging data collection may be essential for some familiar and socially beneficial Internet services and applications. Search engines are one example. Search engines gather detailed data about the contents and structure of the World Wide Web. Consumers understand and depend on search engines to collect this broad range of data and make it available for a wide range of end uses. Search engines also log search queries to improve their services. Search engines may collect such data, which includes personal data, in a manner that is consistent with the Focused Collection principle, so long as their purposes for collecting personal data are clear, and they do not retain personal data beyond the time they need it to achieve any of these purposes.

7.  **ACCOUNTABILITY: Consumers have a right to have personal data handled by companies with appropriate measures in place to assure they adhere to the Consumer Privacy Bill of Rights.** Companies should be accountable to enforcement authorities and consumers for adhering to these principles. Companies also should hold employees responsible for adhering to these principles. To achieve this end, companies should train their employees as appropriate to handle personal data consistently with these principles and regularly evaluate their performance in this regard. Where appropriate, companies should conduct full audits. Companies that disclose personal data to third parties should at a minimum ensure that the recipients are under enforceable contractual obligations to adhere to these principles, unless they are required by law to do otherwise.

Privacy protection depends on companies being accountable to consumers as well as to agencies that enforce consumer data privacy protections. The Accountability principle, however, goes beyond external accountability to encompass practices through which companies prevent lapses in their privacy commitments or detect and remedy any lapses that may occur. Companies that can demonstrate that they live up to their privacy commitments have powerful means of maintaining and strengthening consumer trust. A company's own evaluation can prove invaluable to this process. The appropriate evaluation technique, which could be a self-assessment and need not necessarily be a full audit, will depend on the size, complexity, and nature of a company's business, as well as the sensitivity of the data involved. In recent years, chief privacy officers—experts who raise awareness of privacy issues in companies that face rapid changes in technologies, consumer expectations, and regulations—have emerged as a valuable source of guidance and internal evaluation. Chief privacy officers are likely to provide a continuing source of guidance within companies throughout the development of products and services.

To be fully effective, however, companies should link evaluations to the enforcement of pre-established internal expectations; evaluations are not an end in themselves. Audits—whether conducted by the company or by an independent third party—may be appropriate under some circumstances, but they are not always necessary to fulfill the Accountability principle.

Moreover, accountability must attach to data transferred from one company to another. From the perspective of the Consumer Privacy Bill of Rights, the emphasis is not on the disclosures themselves, but on whether a disclosure leads to a use of personal data that is inconsistent within the context of its collection or a consumer's expressed desire to control the data. Thus, if a company transfers personal data to a third party, it remains accountable and thus should hold the recipient accountable—through contracts or other legally enforceable instruments—for using and disclosing the data in ways that are consistent with the Consumer Privacy Bill of Rights.

# III. Implementing the Consumer Privacy Bill of Rights: Multistakeholder Processes to Develop Enforceable Codes of Conduct

Implementing the general principles in the Consumer Privacy Bill of Rights across the wide range of innovative uses of personal data requires a process to establish more specific practices. The Administration encourages individual companies, industry groups, privacy advocates, consumer groups, crime victims, academics, international partners, State Attorneys General, Federal civil and criminal law enforcement representatives, and other relevant groups to participate in multistakeholder processes to develop codes of conduct that implement these general principles.

In consumer data privacy, as in other areas affecting Internet policy, the Administration believes that multistakeholder processes underlie many of the institutions responsible for the Internet's success. This reflects the Administration's abiding commitment to preserving the Internet as an open, decentralized, user-driven platform for communication, innovation, and economic growth.[27]

The Administration supports open, transparent multistakeholder processes because, when appropriately structured, they can provide the flexibility, speed, and decentralization necessary to address Internet policy challenges. A process that is open to a broad range of participants and facilitates their full participation will allow technical experts, companies, advocates, civil and criminal law enforcement representatives responsible for enforcing consumer privacy laws, and academics to work together to find creative solutions to problems. Flexibility in the deliberative process is critical to allowing stakeholders to explore the technical and policy dimensions—which are often intertwined—of Internet policy issues. Moreover, the United States will need to confront a broad, complex, and global set of consumer data privacy issues for decades to come. A process that works efficiently and on a global scale is therefore essential.

Another key advantage of multistakeholder processes is that they can produce solutions in a more timely fashion than regulatory processes and treaty-based organizations. In the Internet standards world, for example, working groups frequently form around a specific problem and make significant progress toward a solution within months, rather than years. These groups frequently function on the basis of consensus and are amenable to the participation of individuals and groups with limited resources. These characteristics lend legitimacy to the groups and their solutions, which in turn can encourage rapid and effective implementation.

---

27. The United States recently joined the other members of the Organisation for Economic Co-operation and Development (OECD) in recognizing the economic and social importance of the Internet. *See* OECD, Communiqué on Principles for Internet Policy-Making, OECD High-Level Meeting on The Internet Economy: Generating Innovation and Growth, June 28-29, 2011, http://www.ntia.doc.gov/legacy/ntiahome/privwhitepaper.html.

Finally, multistakeholder processes do not rely on a single, centralized authority to solve problems. Specific multistakeholder institutions address specific kinds of Internet policy challenges. This kind of specialization not only speeds up the development of solutions but also helps to avoid the duplication of stakeholders' efforts.

Due in part to its reliance on multistakeholder processes, United States Internet policy has generally avoided fragmented, prescriptive, and unpredictable rules that frustrate innovation and undermine consumer trust. The United States has also refrained from adopting legal requirements that prescribe specific technical requirements, which could fragment the global market for information technologies and services and inhibit innovation. Instead, the United States generally defers to the expert bodies that produce Internet technical standards. In addition, the Administration continues its support for Internet policy processes that are open, transparent, and promote cooperation within a legal framework that sets appropriate performance requirements for individuals and companies.

Consumer data privacy issues exemplify the need for multistakeholder processes that develop the practices and technologies necessary to implement general policy principles. Experience in the United States has shown that both companies and consumers benefit when companies commit to the task of innovating privacy practices. In the early days of commercial activity on the Internet (mid-1990s to early 2000s), for example, the Department of Commerce, the FTC, and the White House convened stakeholders to gather information about privacy issues in this rapidly evolving marketplace. These efforts yielded a flexible, voluntary privacy framework that provided meaningful privacy protections while fostering dynamic innovations in technologies and business models.[28]

Even without legislation, the Administration intends to convene and facilitate multistakeholder processes to produce enforceable codes of conduct. In an open forum, stakeholders with an interest in a specific market or business context will work toward consensus on a legally enforceable code of conduct that implements the Consumer Privacy Bill of Rights. Multistakeholder processes are different from traditional agency rulemakings. The Federal Government will work with stakeholders to establish operating procedures for an open, transparent process. Ultimately, however, the stakeholders themselves will control the process and its results. There is no Federal regulation at the end of the process, and codes will not bind any companies unless they choose to adopt them.

The incentive for stakeholders to participate in this process is twofold. Companies will build consumer trust by engaging directly with consumers and other stakeholders during the process. Adopting a code of conduct that stakeholders develop through this process would further build consumer trust. Second, in any enforcement action based on conduct covered by a code, the FTC will consider a company's adherence to a code favorably.

---

28. For example, the combined efforts of the Department of Commerce, FTC, and the White House produced the consumer data privacy framework of notice and choice, which protected privacy in the context of rapidly developing technologies and markets. *See* FTC, *Privacy Online: Fair Information Practices in the Electronic Marketplace: A Federal Trade Commission Report to Congress, at* http://www.ftc.gov/reports/privacy2000/privacy2000.pdf (2000); White House, *Framework for Global Electronic Commerce, at* § 5, http://clinton4.nara.gov/WH/New/Commerce/ (1997); National Telecommunications and Information Administration, *Privacy and the NII: Safeguarding Telecommunications-Related Personal Information* (Oct. 1995), http://www.ntia.doc.gov/legacy/ntiahome/privwhitepaper.html.

## A.  Building on the Successes of Internet Policymaking

The Internet provides several successful examples of the kind of multistakeholder policy development the Administration envisions. Private-sector standards-setting organizations, for example, are at the forefront of setting Internet-related technical standards. Groups such as the Internet Engineering Task Force (IETF) and the World Wide Web Consortium (W3C) use transparent processes to set Internet-related technical standards. These processes are successful, in part, because stakeholders share an interest in developing consensus-based solutions to the underlying challenges. The success of the resulting standards is evident in the constantly growing range of services and applications—as well as the trillions of dollars in global commerce—they support.

Similarly, the Internet Corporation for Assigned Names and Numbers (ICANN), a nonprofit corporation, coordinates the technical management of the domain name system, which maps domain names to unique numerical addresses. ICANN is also a multistakeholder organization that includes representatives from a broad array of interests, including generic top level domain registries, registrars and registrants, country code top level domain registries, the Regional Internet Registries, root server operators, national governments, and Internet users at large. With this structure, ICANN coordinates the technical management of an important function of the Internet—mapping names that people can remember to numerical addresses that computers can use—and does so in a manner that allows for a wide range of stakeholder input.

Government-convened policymaking efforts, such as the Executive Branch-led privacy discussions of the 1990s and early 2000s, continue to be central to advancing consumer data privacy protections in the United States. The framework in this document is a direct result of the Department of Commerce Internet Policy Task Force's extensive engagement with stakeholders—companies, trade groups, privacy advocates, academics, civil and criminal law enforcement representatives, and foreign government officials. In addition, the FTC has encouraged multistakeholder efforts to develop a "Do Not Track" mechanism, which would afford greater consumer control over personal data in the context of online behavioral advertising.

## B.  Defining the Multistakeholder Process for Consumer Data Privacy

The Department of Commerce's National Telecommunications and Information Administration (NTIA) has the necessary authority and expertise, developed through its role in other areas of Internet policy, to convene multistakeholder processes that address consumer data privacy issues.[29] NTIA will lead the Department of Commerce's convening of stakeholders in a deliberative process that develops codes of conduct and allows stakeholders to adapt the codes to protect consumers' privacy as technologies and market conditions change.[30]

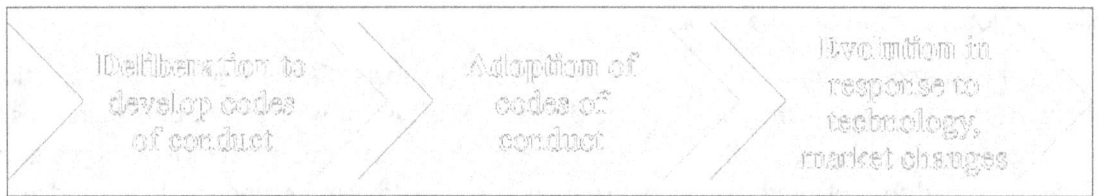

Figure 1. The principal stages of the multistakeholder process for consumer data privacy

### 1.  Deliberation

- **Identifying Issues.** Stakeholder groups, with the assistance of NTIA, will identify markets and industry sectors that involve significant consumer data privacy issues and may be ripe for an enforceable code of conduct. The process will be open, but the focus of a given process likely will not appeal equally to all stakeholders.

- **Initiating and Facilitating Deliberations.** NTIA will take steps to enlist the participation of stakeholders to develop an enforceable code of conduct. As convener, NTIA will open meetings to all stakeholders, including international partners, the FTC, Federal civil and criminal law enforcement representatives, and State Attorneys General, that have an interest in defining an appropriate code of conduct and express a willingness to work in good faith toward reaching consensus on the code's provisions.

  As their first order of business, stakeholders will establish operating processes and procedures. The Administration is committed to a process that is open, transparent, and accommodates participation by groups that have limited resources; however the deliberative process must meet the needs of its participants, who determine and abide by its outcome.[31]

---

29.  NTIA is designated by statute as the "President's principal adviser on telecommunications policies pertaining to the Nation's economic and technological advancement . . . ." 47 U.S.C. § 902(b)(2)(D).

30.  Other Federal agencies may play this convening role if consumer data privacy issues arise in their areas of expertise. Alternatively, private-sector organizations could convene stakeholders, though the dearth of private sector-led code development efforts is precisely the reason that the Administration proposes to serve as convener.

31.  The Administration's guidelines for increasing transparency, participation, and collaboration in public policy development could prove useful here. See President Barack Obama, Memorandum to the Heads of Executive Departments and Agencies: Transparency and Open Government, http://www.whitehouse.gov/the_press_office/TransparencyandOpenGovernment/; Peter R. Orszag, Memorandum for the Heads of Executive Departments and Agencies: Open Government Directive, Dec. 8, 2009, http://www.whitehouse.gov/open/documents/open-government-directive.

- **Conclusion.** A code that reflects the agreement of all stakeholders is ready for companies to consider adopting. The Administration expects, however, that consensus will emerge on parts of a code, and that stakeholders are likely to resolve the most difficult issues later in the process. At this stage, NTIA may need to work intensively with stakeholders to help them resolve their differences. NTIA's role will be to help the parties reach clarity on what their positions are and whether there are options for compromise toward consensus, rather than substituting its own judgment. To minimize the possibility that some stakeholders may draw inflexible lines that prevent consensus, the parties should discuss and set out rules or procedures at the outset of the process to govern how the group will reach an orderly conclusion, even if there is not complete agreement on results.

## 2. Adoption

Once a code of conduct is complete, companies to which the code is relevant may choose to adopt it. The Administration expects that a company's public commitment to adhere to a code of conduct will become enforceable under Section 5 of the FTC Act (15 U.S.C. § 45), just as a company is bound today to follow its privacy statements.[32] Enforceability is essential to assuring consumers that companies' practices match their commitments and thus to strengthening consumer trust.

## 3. Evolution

A key goal of the multistakeholder process is to enable stakeholders to modify privacy protections in response to rapid changes in technology, consumer expectations, and market conditions, to assure they sufficiently protect consumer data privacy. The multistakeholder process offers several ways to keep codes of conduct current. Stakeholders may decide at any time that a code of conduct no longer provides effective consumer data privacy protections, in light of technological or market changes. NTIA might also draw this conclusion and seek to re-convene stakeholders. As with the initial development of a code of conduct, however, stakeholder participation in the process to revise a code of conduct would be voluntary. The Federal Government would not revise a code of conduct; rather, stakeholder groups will make these changes with Federal Government input. Finally, under the legislative safe harbor framework discussed in the following section, Congress could prescribe a renewal period for codes of conduct, so that the FTC periodically reviews codes that are the basis of enforcement safe harbors.

---

32. The FTC brings cases based on violations of commitments in its privacy statements under its authority to prevent deceptive acts or practices. In addition, the FTC brings data privacy cases under its unfairness jurisdiction, which will remain an important source of consumer data privacy protection.

# IV. Building on the FTC's Enforcement Expertise

## A. Protecting Consumers Through Strong Enforcement

Enforcement is critical to ensuring that the privacy commitments companies make by adopting a code of conduct are meaningful. Self-regulatory bodies, which develop and administer voluntary guidelines for member companies, can provide a first line of enforcement, though they are not necessary for the framework described here. Enforcement through self-regulatory bodies can help to detect and remedy compliance issues at an early stage. As a result, this kind of enforcement can strengthen trust in a code of conduct and the companies that commit to the code.

Government agencies also play a vital role in enforcing the privacy protections in codes of conduct. The FTC is the Federal Government's leading consumer privacy enforcement authority.[33] Enforcement actions by the FTC (and State Attorneys General) have established that companies' failures to adhere to voluntary privacy commitments, such as those stated in privacy policies, are actionable under the FTC Act's (and State analogues) prohibition on unfair or deceptive acts or practices.[34] In addition, the FTC brings cases against companies that allegedly failed to use reasonable security measures to protect personal information about consumers.[35] Using this authority, the FTC has brought cases that effectively protect consumer data privacy within a flexible and evolving approach to changing technologies and markets. The same authority would allow the FTC to enforce the commitments of companies under its jurisdiction to adhere to codes of conduct developed through the multistakeholder process.[36] Thus, companies that adopt codes of conduct will make commitments that are legally enforceable under existing law.

## B. Providing Incentives to Develop Enforceable Codes of Conduct

The FTC has significant enforcement and policy expertise to offer all stakeholders on consumer data privacy issues codes of conduct. With or without consumer data privacy legislation, the FTC should provide assistance and advice regarding development of the codes. In the absence of legislation, the FTC, Federal civil and criminal law enforcement representatives, and States should participate in the multistakeholder deliberations by providing advice on substance and process. Once stakeholders have developed a code, a company may voluntarily adhere to the code in order to gain greater certainty and

---

33. Note, however, the FTC does not currently have authority to enforce Section 5 of the FTC Act, 15, U.S.C. § 45, against certain corporations that operate for profit.

34. *See* FTC Act § 5, 15 U.S.C. § 45. In addition to using its Section 5 authority to protect consumer data privacy, the FTC has brought dozens of cases under sector-specific statutes, such as the Children's Online Privacy Protection Act, the Fair Credit Reporting Act, the Gramm-Leach-Bliley Act, and the Do Not Call Rule. For a review of these cases, see FTC Staff Report at 9-13.

35. *See* FTC Staff Report at 10 (reviewing enforcement actions that include counts based on unfair acts or practices).

36. The FTC's jurisdiction over nonprofits and certain other types of entities under FTC Act § 5 may be limited.

assure its customers that its practices protect their privacy. Companies may choose to adopt multiple codes of conduct to cover different lines of business; the common baseline of the Consumer Privacy Bill of Rights should help ensure that the codes are consistent. Then, in any investigation or enforcement action related to the subject matter of one or more codes, the FTC should consider the company's adherence to the codes favorably.

# V. Promoting International Interoperability

The Internet helps U.S. companies expand across borders. As a result, cross-border data flows are a vital component of the domestic and global economies. Differences in national privacy laws create challenges for companies wishing to transfer personal data across national borders. Complying with different privacy laws is burdensome for companies that transfer personal data as part of well-defined, discrete data processing operations because legal standards may vary among jurisdictions, and companies may need to obtain multiple regulatory approvals to conduct even routine operations.

Services that cater to individual users face steeper compliance challenges because they handle data flows that are more complex and varied. Further complicating matters is the proliferation of cloud computing systems.[37] This globally distributed architecture helps deliver cost-effective, innovative new services to consumers, companies, and governments. It also allows consumers and companies to send the personal data they generate and use to recipients all over the world. Consumer data privacy frameworks should not only facilitate these technologies and business models but also adapt rapidly to those that have yet to emerge.

Though governments may take different approaches to meeting these challenges, it is critical to the continued growth of the digital economy that they strive to create interoperability between privacy regimes. The Administration believes flexible multistakeholder processes that address novel uses and transfers of data facilitate interoperable privacy regimes. The United States is committed to engaging with its international partners to increase interoperability in privacy laws by pursuing mutual recognition, the development of codes of conduct through multistakeholder processes, and enforcement cooperation. It is also committed to including international counterparts in these multistakeholder processes, to enable global consensus on emerging privacy issues.

## A.   Mutual Recognition

Mutual recognition of commercial data privacy frameworks is a means to achieve meaningful global data protection. A starting point for mutual recognition is the embrace of common values surrounding privacy and personal data protection. Two principles should determine whether the conditions for mutual recognition between specific privacy frameworks exist: effective enforcement and mechanisms that allow companies to demonstrate accountability.

Where companies are under comparable legal requirements, mutual recognition means that all parties can enforce the companies' obligations. Effective enforcement, conducted according to publicly announced policies, is therefore critical to establishing interoperability. Enforcement authorities and mechanisms vary from country to country, and the United States recognizes that a variety of approaches can be effective. The United States relies primarily upon the FTC's case-by-case enforcement of general

---

37.   NIST has identified five essential characteristics of cloud computing: on-demand self-service, broad network access, resource pooling, rapid elasticity, and measured service. *See supra* note 6.

prohibitions on unfair or deceptive acts and practices. This approach helps develop evolving standards for handling personal data in the private sector.

In the context of mutual recognition, accountability refers to a company's capacity to demonstrate the implementation of enforceable policies and procedures relating to privacy (whether adopted voluntarily or as a result of legal obligations). Accountability mechanisms include self-assessments, evaluations, and audits.[38] The Administration encourages stakeholders to work together to identify globally accepted accountability mechanisms when developing codes of conduct.

One example of an initiative to facilitate transnational mutual recognition is the Asia-Pacific Economic Cooperation's (APEC) voluntary system of Cross Border Privacy Rules (CBPR), which is based on the APEC Privacy Framework and includes privacy principles that APEC member economies have agreed to recognize.[39] Codes of conduct based on these principles could streamline the data privacy policies and practices of companies operating throughout the vast APEC region.[40] Upon implementation, APEC's CBPR system will require interested applicants to demonstrate that they comply with a set of CBPR program requirements based on the APEC Privacy Framework. Moreover, the commitments an applicant makes during this process, while voluntary, must be enforceable under laws in member economies. Successful CBPR certification will entitle participating companies to represent to consumers that they are accountable and meet stringent and globally recognized standards, thereby facilitating the transfer of personal data throughout the APEC region.

In Europe, Article 27 of European Union (EU) Directive 95/46/EC on the protection of individuals with regard to the processing of personal data and on the free movement of such data, commonly known as the EU Data Protection Directive, encourages the development of codes of conduct to help implement the law. Like the Administration's framework, which proposes industry-specific codes of conduct, the Data Protection Directive recognizes that codes of conduct that implement general privacy principles may differ in their details, according to the needs of the relevant industry. The Administration is committed to working with organizations at the EU level as well as with member states to make codes of conduct the basis of mutually recognized privacy protections.

The Safe Harbor Frameworks that the United States developed with the EU and Switzerland are early examples of global interoperability that have had a meaningful impact on transatlantic data flows. The United States, the EU, and Switzerland negotiated these Frameworks to accomplish the objectives of protecting personal information while also ensuring that companies could transfer information in a way that did not disrupt their global business operations. These Frameworks allow companies to self-certify that they comply with requirements under the EU Data Protection Directive, subject to FTC

---

38. Auditing is not a requirement under the Accountability principle stated in the Consumer Privacy Bill of Rights. This section discusses the potential use of audits by companies that seek to take advantage of global interoperability in privacy laws. Not all organizations, however, fit this description.

39. The nine principles are collection limitation, integrity of personal information, notice, uses of personal information, choice, security safeguards, access and correction, accountability, and harm prevention. *See* http://publications.apec.org/publication-detail.php?pub_id=390.

40. Currently, APEC includes 21 members: Australia, Brunei Darussalam, Canada, Chile, the People's Republic of China, Hong Kong, Indonesia, Japan, the Republic of Korea, Malaysia, Mexico, New Zealand, Papua New Guinea, Peru, the Philippines, Russia, Singapore, Chinese Taipei, Thailand, the United States, and Vietnam. APEC, Member Economies, http://www.apec.org/About-Us/About-APEC/Member-Economies.aspx (last visited Sept. 7, 2011).

enforcement of these representations.[41] The more than 2,700 companies that participate in the Safe Harbor Frameworks may transfer personal data from the EU to the United States. As a result, the Safe Harbor Frameworks have effectively reduced barriers to personal data flow and thereby support trade and economic growth.

## B.   An International Role for Multistakeholder Processes and Codes of Conduct

The attributes of speed, flexibility and decentralized problem-solving in well-structured multistake-holder consultations offer certain advantages over traditional government regulation when it comes to establishing globally applicable rules and guidelines that promote innovation and protect consumers. Multistakeholder-developed codes of conduct, combined with existing mutual recognition frameworks, hold the promise of greatly simplifying companies' compliance burdens.

While the Safe Harbor Frameworks have proven to be valuable in facilitating transatlantic trade, they are not perfect solutions for all U.S. entities. Sectors not regulated by the FTC, such as financial services, telecommunications common carriers, and insurance, are not covered by the Safe Harbor Frameworks. Some companies in these sectors have indicated that they would like to see an improved environment for transatlantic data transfers.

To build on the success of the Safe Harbor Frameworks, the Administration, through the Departments of Commerce and State, plans to develop additional mechanisms—such as jointly developed codes of conduct—that support mutual recognition of legal regimes, facilitate the free flow of information, and address emerging privacy challenges. The Administration hopes to include international stakeholders in the multistakeholder processes. The Safe Harbor Frameworks could one day be supplemented by codes of conduct reflecting transatlantic consensus on important, emerging privacy issues.

## C.   Enforcement Cooperation

To realize global interoperability in data protection, mutual recognition must be accompanied by robust enforcement cooperation. Such collaboration, whether bilateral or multilateral, is necessary to address information sharing among data protection authorities.

Empowered by legislation that grants it greater authority to cooperate with foreign counterparts, the FTC helped to create the Global Privacy Enforcement Network ("GPEN"). GPEN aims to further the development of privacy enforcement priorities, sharing of best practices, and support for joint enforcement initiatives. The FTC is involved in a number of other international organizations, including the OECD, APEC, the Asia-Pacific Privacy Authorities forum, and the International Conference of Data Protection and Privacy Commissioners. The work of the United States Government in GPEN, the OECD, APEC, and other venues is increasing collaboration in privacy investigations and enforcement actions globally. Given that Internet-based services reach individuals in jurisdictions around the world, it is neither effective nor wise policy for governments to enforce national data privacy legislation in isolation.

---

41.   For a summary of the FTC's enforcement of the U.S.-EU Safe Harbor Framework, see FTC, FTC Settles with Six Companies Claiming to Comply with International Privacy Framework, Oct. 6, 2009, http://www.ftc.gov/opa/2009/10/safeharbor.shtm. *See also In re* Google, Inc., Complaint, at 7 File No. 102 3136, Mar. 30, 2011 (alleging "respondent did not adhere to the US Safe Harbor Privacy Principles of Notice and Choice").

# VI. Enacting Consumer Data Privacy Legislation

The Administration urges Congress to pass legislation adopting the Consumer Privacy Bill of Rights. Legislation would promote trust in the digital economy by providing a basic set of privacy rights throughout areas of the commercial sector that are not currently subject to specific Federal data privacy legislation. The flexible approach that the Administration supports will allow companies to implement the Consumer Privacy Bill of Rights in ways that fit the context in which they do business.

## A. Codify the Consumer Privacy Bill of Rights

Congress should act to protect consumers from violations of the rights defined in the Administration's proposed Consumer Privacy Bill of Rights. These rights provide clear protection for consumers and define rules of the road for the rapidly growing marketplace for personal data.[42] The legislation should permit the FTC and State Attorneys General to enforce these rights directly. The legislation will need to state companies' obligations under the Consumer Privacy Bill of Rights with greater specificity than this document provides. The Consumer Privacy Bill of Rights is a guide for the Administration to work collaboratively with Congress on statutory language.[43]

To provide greater legal certainty and to encourage the development and adoption of industry-specific codes of conduct, the Administration also supports legislation that authorizes the FTC to review codes of conduct and grant companies that commit to adhere—and do adhere—to such codes forbearance from enforcement of provisions of the legislation.

In addition, consumer data privacy legislation should avoid:

- Adding duplicative or overly burdensome regulatory requirements to companies that are already adhering to legislatively adopted privacy principles.

- Prescribing technology-specific means of complying with the law's obligations.

- Precluding new business models that are consistent with the Consumer Privacy Bill of Rights in general but may involve new uses of personal information not contemplated at the time the statute is written.

- Altering existing statutory or regulatory authorities pursuant to which the government may obtain information that is necessary to assist in conducting border searches, investigating criminal conduct or other violations of law, or protecting public safety and national security.

---

42. The Administration is separately considering the need to amend laws pertaining to the government's access to data in the possession of private parties, including the Electronic Communications Privacy Act, to address changes in technology.

43. In the absence of legislation, the Consumer Privacy Bill of Rights set forth in this document provides guidance for stakeholders and does not alter the FTC's existing enforcement authority under FTC Act § 5.

- Contravening the ability of law enforcement to investigate and prosecute criminal acts, and ensure public safety.

- Altering existing statutory, regulatory, or policy authorities that apply to the government's information practices or address privacy issues outside of a purely commercial, consumer-oriented context.

## B.  Grant the FTC Direct Enforcement Authority

The Administration encourages Congress to grant the FTC the authority to enforce each element of the statutory Consumer Privacy Bill of Rights.[44] This authority would provide greater certainty to consumers and companies both. Companies would begin with a clearer roadmap to their privacy obligations. Consumers would benefit from knowing that Congress has empowered the FTC to enforce a comprehensive set of privacy protections in the commercial marketplace. At the same time, a statute that allows the FTC to enforce the Consumer Privacy Bill of Rights directly would provide flexibility and permit the FTC to address emerging privacy issues through specific enforcement actions governed by applicable procedural safeguards. Companies seeking even greater certainty under such legislation should use the multistakeholder process and enforcement safe harbor discussed below to develop context-specific codes of conduct in a timely fashion. The Administration recommends that Congress grant the same authority to State Attorneys General. So long as they coordinate with the FTC in their enforcement actions, States could provide additional enforcement resources and a considerable source of consumer data privacy expertise.

In domains involving rapid changes in technology and business practices, Congress has chosen to create flexible standards rather than tailoring them to technologies and practices that exist at the time it passes a law. In the realm of antitrust, for example, the Sherman Act prohibits agreements "in restraint of trade."[45] The Copyright Act defines basic terms such as "copies," "devices," and "processes" with reference to technologies "now known or later developed."[46] And, in the realm of data privacy, the FTC has brought numerous enforcement actions under the FTC Act Section 5's prohibition on "unfair or deceptive acts or practices." A combination of agency guidelines, judicial interpretation, and industry practices provides interpretations of these terms to allow individuals and companies to determine with greater certainty whether their conduct complies with these general laws.

The Administration encourages Congress to follow a similar path with baseline consumer data privacy legislation. It is important that a baseline statute provide a level playing field for companies, a consistent set of expectations for consumers, and greater clarity and transparency in the basis for FTC enforcement actions. The FTC also could engage the public to clarify how it will enforce the statutory Consumer Privacy Bill of Rights. The primary mechanisms to clarify the statute's requirements should be the multistakeholder process and enforcement safe harbor, based on enforceable codes of conduct, as discussed below. The more traditional modes of clarifying general statutory requirements, however, could also play a helpful role.

---

44.  The FTC refers civil penalty actions to the Department of Justice, which may bring an action within 45 days. If the Department of Justice declines to litigate, the FTC may prosecute the case itself. *See, e.g.,* 15 U.S.C. § 56(a).

45.  15 U.S.C § 1.

46.  17 U.S.C. § 101.

## C.   Provide Legal Certainty Through an Enforcement Safe Harbor

The Administration supports authorizing the FTC to provide greater assurance to companies that adopt enforceable codes of conduct than is possible under current law. Two legislative structures would help to accomplish this goal. First, the FTC should have explicit authority to review codes of conduct against the Consumer Privacy Bill of Rights, as they are set forth in legislation. Legislation should require the FTC to review codes submitted for review within a reasonable amount of time (e.g., 180 days), require the FTC to consider public comments on a code, limit its review authority to approving or rejecting a code that reflects the consensus of all participants in the multistakeholder process, and establish a period for reviewing approved codes to ensure that they sufficiently protect consumer privacy in light of technological and market changes. The record from the multistakeholder process that produced a code—and particularly the presence of general consensus on its provisions—would help to guide the FTC's assessment of whether a code sufficiently implements the Consumer Privacy Bill of Rights. Because the outcome of FTC review will likely influence companies' decisions to adopt codes of conduct—the end result of the multistakeholder process—it is appropriate to determine the details of FTC review through a process that is open to all stakeholders. These details, however, need to be legally binding. Accordingly, the Administration recommends that Congress grant the FTC authority under the Administrative Procedure Act (5 U.S.C. § 552 *et seq.*) to issue rules that establish a fair and transparent process for reviewing and approving codes of conduct.

The second element that the Administration recommends is giving the FTC the authority to grant a "safe harbor"—that is, forbearance from enforcement of the statutory Consumer Privacy Bill of Rights—to companies that follow a code of conduct that the FTC has reviewed and approved. Companies that decline to adopt a code of conduct, or choose not to seek FTC review of a code that they do adopt, would simply be subject to the general obligations of the legislatively adopted Consumer Privacy Bill of Rights.

## D.   Balance Federal and State Roles in Consumer Data Privacy Protection

Federal legislation that enacts a Consumer Privacy Bill of Rights should provide a national standard for protecting consumer data privacy where existing Federal data privacy statutes do not apply. Nationally uniform consumer data privacy rules are necessary to create certainty for companies and consistent protections for consumers. These rules should take into consideration the need for certain information to be available for law enforcement-related purposes. Moreover, national uniformity is crucial to preserving the incentives that the Administration's framework provides through the multistakeholder process. Stakeholders' incentives to participate in the multistakeholder process, and companies' incentives to adopt codes of conduct, would be diminished if States enacted laws with more stringent requirements. The Administration therefore recommends that Congress preempt State laws to the extent they are inconsistent with the Consumer Privacy Bill of Rights as enacted and applied. The Administration also recommends that Congress provide forbearance from enforcement of State laws against companies that adopt and comply with FTC-approved codes of conduct.

The Administration's proposed approach preserves important policymaking and enforcement roles for the States. States can and should play a highly constructive role in the multistakeholder process. The Administration also supports granting State Attorneys General with the authority to enforce the

Consumer Privacy Bill of Rights. Taken together, these mechanisms will provide States means to address consumer data privacy issues that States identify while maintaining uniformity at the national level. The Administration will also work with Congress, States, the private sector, and other stakeholders to determine whether there are specific sectors in which States could enact laws that would not disrupt the broader uniformity the Administration seeks in consumer data privacy protections. For example, it may be appropriate to allow States to enact laws that apply the Consumer Privacy Bill of Rights to personal data in sectors they closely regulate, such as retail electricity distribution.[47]

## E.  Preserve Effective Protections in Existing Federal Data Privacy Laws

Consumer data privacy legislation should preserve existing sector-specific Federal laws that effectively protect personal data, minimize the duplication of legal requirements, and provide consumers with a clear sense of what protections they have and who enforces them. Where existing Federal laws do not meet these guidelines, however, the Administration encourages Congress to consider how consumer data privacy legislation could simplify existing requirements, to the benefit of consumers and companies.

In general, the sector-specific Federal data privacy laws establish legal obligations that are tailored to the sensitivity of personal data used and the prevailing practices in those sectors.[48] For instance, HIPAA and the HIPAA Privacy and Security Rules regulate the collection, use, and disclosure of personal health information by healthcare providers, insurers, and health information clearinghouses. HIPAA permits by default personal health information practices that are necessary or commonly accepted in the healthcare context, such as disclosures of personal health information between two healthcare providers in order to treat a patient. Federal data privacy laws that apply to education, credit reporting, financial services, and the collection of children's personal data are examples of similarly well-tailored requirements.

### 1.  Create Comprehensive Privacy Protection Without Duplicating Burdens

To avoid creating duplicative regulatory burdens, the Administration supports exempting companies from consumer data privacy legislation to the extent that their activities are subject to existing Federal data privacy laws. However, activities within such companies that do not fall under an existing data privacy law would be covered by the legislation that the Administration proposes. The alternative— exempting entire entities that are subject to an existing Federal data privacy law—could allow the exception to swallow the rule. For example, the Gramm-Leach-Bliley Act (GLB) requires financial institutions to take certain privacy and security precautions with nonpublic personal information. If entities that are subject to GLB were exempt from a baseline consumer data privacy law for non-GLB-covered personal data, the baseline statute's effectiveness could be significantly diminished.

---

47.   Indeed, the Administration recently called for State public utilities commissions to follow privacy principles that are very similar to those in the Consumer Privacy Bill of Rights in order to protect personal data associated with the "smart" electric grid. *See supra* note 23.

48.   This limitation also means that the laws that regulate the Federal government's collection, use, and disclosure of personal data are beyond the framework's scope.

## 2.   *Amend Laws That Create Inconsistent or Confusing Requirements*

Because existing Federal laws treat similar technologies within the communications sector differently,[49] the Administration supports simplifying and clarifying the legal landscape and making the FTC responsible for enforcing the Consumer Privacy Bill of Rights against communications providers.

## F.   Set a National Standard for Security Breach Notification

In the specific area of security breaches, the Administration supports creating a national standard under which companies must notify consumers of unauthorized disclosures of certain kinds of personal data. Security breach notification (SBN) laws effectively promote the protection of sensitive personal data. They require companies in certain situations to notify consumers whose personal data was exposed to unauthorized recipients. Notice helps consumers protect themselves against harms such as identity theft. It also provides companies with incentives to establish better data security in the first place. The SBN model is also gaining acceptance internationally as a performance-based requirement that effectively protects consumers.

Currently, 47 States, the District of Columbia, and several U.S. Territories, have SBN laws. Variations in States have allowed a sense of the most effective approaches to emerge, but the need for national uniformity is now evident. The patchwork of State laws creates significant burdens for companies without much countervailing benefit for consumers. As part of its comprehensive cybersecurity legislative package, the Administration recommended creating a national standard for notifying consumers in the event that there are unauthorized disclosures of certain types of personal data.[50] This national standard would replace the various State standards that exist today and preempt future State legislation in this area.

---

49.   *See*, e.g., 47 U.S.C. §§ 222, 338 & 551 (requiring telecommunications carriers, satellite carriers, and cable services, respectively, to protect customers' personal information).

50.   The White House, Data Breach Notification Legislative Language, May 2011, http://www.whitehouse.gov/sites/default/files/omb/legislative/letters/data-breach-notification.pdf.

# VII. Federal Government Leadership in Improving Individual Privacy Protections

In areas other than consumer data privacy, the Administration is continuing the Federal government's long history of championing data privacy protections in the public and private spheres. This history stems from the early days of computerized data processing. In 1973, the Department of Health, Education, and Welfare (HEW) Advisory Committee on Automated Personal Data Systems issued a report entitled Records, Computers, and the Rights of Citizens. This landmark report provided an early statement of the FIPPs that provide a foundation for the Administration's Consumer Privacy Bill of Rights.

Since then, the Federal government has led the way in demonstrating that protecting privacy is integral to conducting the Nation's business. No single event or policy need has spurred this activity. In some cases, Federal agencies consider privacy issues in response to specific Congressional mandates. In other cases, Federal agencies integrate privacy into innovative initiatives that advance their core missions. The activities of Federal agencies with duties that range across a broad array of economic sectors—including healthcare, financial services, and education—illustrate the Administration's commitment to promoting best practices, enabling new services, providing tools to address many different privacy issues, and enforcing individual privacy rights.

## A. Enabling New Services

Like the private sector, Federal agencies must confront data privacy issues when delivering services to the public. A particularly challenging set of privacy issues arises in connection with delivering healthcare to the Nation's veterans. The Department of Veterans Affairs (VA) provides healthcare for 8.3 million enrolled veterans through more than 1,400 facilities distributed across the Nation. To help manage a healthcare operation of this scale and scope efficiently and cost-effectively, the VA is continuing to incorporate information technology into its healthcare delivery system. Protecting the privacy of veterans' health information is essential to the success of this endeavor.

VA recently launched an initiative that demonstrates how careful attention to privacy and security protections for personal health information can lead to significant advances in how healthcare is delivered. VA incorporated privacy and security protections into its "My HealtheVet Personal Health Record." This system is a gateway to information that helps veterans to enable their caregivers to deliver better care and provides other Internet-based tools that empower veterans to become active partners in their health care. The VA's Blue Button service allows veterans to download an electronic copy of their HealtheVet information in a secure manner.

**How Administration Action Is Enabling Privacy in Other Areas**

- **Integrating Privacy into Cybersecurity Initiatives.** Protecting privacy is a priority in the Administration's efforts to secure online environments for continuing increases in productivity, innovation, and support for new business ventures. Led by the National Institute of Standards and Technology (NIST), the *National Strategy for Trusted Identities in Cyberspace* calls for a partnership with the commercial sector to develop more standardized, secure, and privacy-enhancing ways to authenticate individuals online.

- **Enhancing Transparency in Credit Markets.** The Administration is ensuring that privacy protections keep pace with developments in uses of personal data in setting the terms of consumer credit. The Federal Reserve Board, together with the FTC, issued a rule that requires creditors to provide a consumer with notice when, based on the consumer's credit report, the creditor provides credit to the consumer on less favorable terms than it provides to other consumers. This rule also entitles consumers who are notified of such "risk-based pricing" to obtain a free credit report, so that they can check whether the information creditors use is accurate.

## B. Protecting Privacy Through Effective Enforcement

The FTC has used its civil enforcement authority against those commercial enterprises that fail to follow Commission rules or act in an unfair or deceptive manner. Since 2009, the FTC has taken actions against companies that have failed to exercise reasonable care to secure sensitive personal and medical information, represented that they abide by the U.S.-EU or U.S.-Swiss Safe Harbor agreements when they do not or they have allowed these certifications to lapse, or that misrepresent the use of tracking software. The FTC also prosecuted actions involving deceptive practices by online seal providers, social media companies, and companies claiming to protect identities. In addition, the FTC prosecuted cases under the Telemarketing Sales Rule, the COPPA Rule, the Fair Credit Reporting Act, and the GLB Safeguards Rule.

The Administration also takes enforcing statutory privacy rights seriously. Federal agencies with law enforcement authority have taken action against those who violate privacy rights. For example, the Department of Justice (DOJ) aggressively prosecutes cases involving identity theft—the use of misappropriated personal data that can cause life-disrupting and economically devastating harm to its victims. In 2010 alone, DOJ's United States Attorneys' Offices prosecuted nearly 1300 cases involving identity theft, and U.S. Attorneys have brought nearly 700 identity theft cases in the current fiscal year. DOJ, assisted by investigators from the Federal Bureau of Investigation and Department of Homeland Security (DHS) components such as United States Secret Service and U.S. Immigration and Customs Enforcement, also vigorously prosecutes individuals who obtain personal data (and other information) by breaking into computers. Taken together, these efforts help protect the confidentiality of personal data and bring justice for victims of identity theft and other crimes that involve the misuse of personal data.

## C.  Guidance for Protecting Privacy

Federal agencies are also devoting resources to producing guidance on data privacy that has broad applicability in the private sector. The Department of Health and Human Services (HHS), for example, has issued guidance that analyzes some of the fundamental issues surrounding responses to security breaches that involve personally identifiable information. In 2009, the Department of Health and Human Services Office for Civil Rights (OCR) issued guidance on when health information is considered to be secure (and therefore exempt from breach notification requirements) by specifying the technologies and methodologies that render protected health information unusable, unreadable, or indecipherable. In 2010, OCR also issued guidance on conducting a risk analysis under the HIPAA Security Rule. OCR plans to issue additional guidance on the HIPAA Privacy Rule's "minimum necessary" standard and on de-identification of health information under the HIPAA Privacy Rule.

Federal agencies are also providing guidance on how to make more effective use of existing privacy-protecting measures. In 2009, eight Federal agencies released a model privacy notice form that financial institutions can opt to use for their privacy notices to consumers required by GLB. Use of the model form provides a legal safe harbor for compliance with the GLB Privacy Rule, though the model form is not required. The agencies conducted extensive consumer research and testing in developing the model form to ensure that consumers can easily understand what financial institutions do with their personal information and compare different institutions' information sharing practices.

**Other Significant Administration Guidance on Privacy:**

- **Raising Public Awareness of Privacy and Data Security.** DHS is leading a national public awareness effort called *Stop. Think. Connect.* to inform the American public of the need to strengthen cybersecurity and to provide practical tips to help Americans increase their safety and security online. In addition, the FTC has issued guides explaining measures that consumers and companies can take to protect children's privacy online, minimize the risk of medical identity theft, and prevent the loss of sensitive data through peer-to-peer file sharing applications.

- **Applying Privacy Principles to New Technologies.** The Administration is demonstrating that the same privacy principles that inform the general consumer data privacy framework developed here also apply to specific, emerging contexts. The "Smart Grid"—the incorporation of information technologies to make the electric grid more efficient, more accommodating of clean sources of energy, and a source of new jobs and innovation—provides an excellent example. Over the past two years, the Department of Energy and the National Institute of Standards and Technology engaged with stakeholders to understand privacy issues that could arise from this promising new technology. This work culminated in the Administration's *Policy Framework for The 21st Century Grid: Enabling Our Secure Energy Future*, which recommends that States make comprehensive FIPPs the starting point for protecting the detailed energy usage data that the Smart Grid will generate.

## D.  Integrating Privacy Into the Structure of Federal Agencies

Finally, Federal agencies are leading the way in incorporating privacy into their structure and operations and in developing accountable organizations. Some of these accountability-enhancing practices and tools have diffused to the private sector and across the globe. For example, the Internal Revenue Service and DHS pioneered the use of privacy impact assessments (PIAs), which provide for structured assessments of the potential privacy issues arising from new information systems and, under the E-Government Act of 2002, are now required of Federal agencies under some circumstances. Building on efforts of previous Administrations, this Administration has extended the use of PIAs to social media. Since their initial development within the Federal government, PIAs have become widely used in the private sector and within the European Union. Federal agencies also continue to make privacy professionals part of their senior leadership structures. Many Federal agencies have full-time, professional chief privacy officers, who engage on privacy issues within their agencies, in broader discussions within the Federal government, and with the general public.

# VIII. Conclusion

The United States is committed to protecting privacy. It is an element of individual dignity and an aspect of participation in democratic society. To an increasing extent, privacy protections have become critical to the information-based economy. Stronger consumer data privacy protections will buttress the trust that is necessary to promote the full economic, social, and political uses of networked technologies. The increasing quantities of personal data that these technologies subject to collection, use, and disclosure have fueled innovation and significant social benefits. We can preserve these benefits while also ensuring that our consumer data privacy policy better reflects the value that Americans place on privacy and bolsters trust in the Internet and other networked technologies.

The framework set forth in the preceding pages provides a way to achieve these goals. The Consumer Privacy Bill of Rights should be the legal baseline that governs consumer data privacy in the United States. The Administration will work with Congress to bring this about, but it will also work with private-sector stakeholders to adopt the Consumer Privacy Bill of Rights in the absence of legislation. To encourage adoption, the Department of Commerce will convene multistakeholder processes to encourage the development of enforceable, context-specific codes of conduct. The United States Government will engage with our international partners to increase the interoperability of our respective consumer data privacy frameworks. Federal agencies will continue to develop innovative privacy-protecting programs and guidance as well as enforce the broad array of existing Federal laws that protect consumer privacy.

A cornerstone of this framework is its call for the ongoing participation of private-sector stakeholders. The views that companies, civil society, academics, and advocates provided to the Administration through written comments, public symposia, and informal discussions have been invaluable in shaping this framework. Implementing it, and making progress toward consumer data privacy protections that support a more trustworthy networked world, will require all of us to continue to work together.

# Appendix A: The Consumer Privacy Bill of Rights

## CONSUMER PRIVACY BILL OF RIGHTS

The Consumer Privacy Bill of Rights applies to *personal data*, which means any data, including aggregations of data, which is linkable to a specific individual. Personal data may include data that is linked to a specific computer or other device. The Administration supports Federal legislation that adopts the principles of the Consumer Privacy Bill of Rights. Even without legislation, the Administration will convene multistakeholder processes that use these rights as a template for codes of conduct that are enforceable by the Federal Trade Commission. These elements—the Consumer Privacy Bill of Rights, codes of conduct, and strong enforcement—will increase interoperability between the U.S. consumer data privacy framework and those of our international partners.

1.  **INDIVIDUAL CONTROL: Consumers have a right to exercise control over what personal data companies collect from them and how they use it.** Companies should provide consumers appropriate control over the personal data that consumers share with others and over how companies collect, use, or disclose personal data. Companies should enable these choices by providing consumers with easily used and accessible mechanisms that reflect the scale, scope, and sensitivity of the personal data that they collect, use, or disclose, as well as the sensitivity of the uses they make of personal data. Companies should offer consumers clear and simple choices, presented at times and in ways that enable consumers to make meaningful decisions about personal data collection, use, and disclosure. Companies should offer consumers means to withdraw or limit consent that are as accessible and easily used as the methods for granting consent in the first place.

2.  **TRANSPARENCY: Consumers have a right to easily understandable and accessible information about privacy and security practices.** At times and in places that are most useful to enabling consumers to gain a meaningful understanding of privacy risks and the ability to exercise Individual Control, companies should provide clear descriptions of what personal data they collect, why they need the data, how they will use it, when they will delete the data or de-identify it from consumers, and whether and for what purposes they may share personal data with third parties.

3.  **RESPECT FOR CONTEXT: Consumers have a right to expect that companies will collect, use, and disclose personal data in ways that are consistent with the context in which consumers provide the data.** Companies should limit their use and disclosure of personal data to those purposes that are consistent with both the relationship that they have with consumers and the context in which consumers originally disclosed the data, unless required by law to do otherwise. If companies will use or disclose personal data for other purposes, they should provide heightened Transparency and Individual Control by disclosing these other purposes in a manner that is prominent and easily actionable by consumers at the time of data collection. If,

subsequent to collection, companies decide to use or disclose personal data for purposes that are inconsistent with the context in which the data was disclosed, they must provide heightened measures of Transparency and Individual Choice. Finally, the age and familiarity with technology of consumers who engage with a company are important elements of context. Companies should fulfill the obligations under this principle in ways that are appropriate for the age and sophistication of consumers. In particular, the principles in the Consumer Privacy Bill of Rights may require greater protections for personal data obtained from children and teenagers than for adults.

4.  **SECURITY: Consumers have a right to secure and responsible handling of personal data.** Companies should assess the privacy and security risks associated with their personal data practices and maintain reasonable safeguards to control risks such as loss; unauthorized access, use, destruction, or modification; and improper disclosure.

5.  **ACCESS AND ACCURACY: Consumers have a right to access and correct personal data in usable formats, in a manner that is appropriate to the sensitivity of the data and the risk of adverse consequences to consumers if the data is inaccurate.** Companies should use reasonable measures to ensure they maintain accurate personal data. Companies also should provide consumers with reasonable access to personal data that they collect or maintain about them, as well as the appropriate means and opportunity to correct inaccurate data or request its deletion or use limitation. Companies that handle personal data should construe this principle in a manner consistent with freedom of expression and freedom of the press. In determining what measures they may use to maintain accuracy and to provide access, correction, deletion, or suppression capabilities to consumers, companies may also consider the scale, scope, and sensitivity of the personal data that they collect or maintain and the likelihood that its use may expose consumers to financial, physical, or other material harm.

6.  **FOCUSED COLLECTION: Consumers have a right to reasonable limits on the personal data that companies collect and retain.** Companies should collect only as much personal data as they need to accomplish purposes specified under the Respect for Context principle. Companies should securely dispose of or de-identify personal data once they no longer need it, unless they are under a legal obligation to do otherwise.

7.  **ACCOUNTABILITY: Consumers have a right to have personal data handled by companies with appropriate measures in place to assure they adhere to the Consumer Privacy Bill of Rights.** Companies should be accountable to enforcement authorities and consumers for adhering to these principles. Companies also should hold employees responsible for adhering to these principles. To achieve this end, companies should train their employees as appropriate to handle personal data consistently with these principles and regularly evaluate their performance in this regard. Where appropriate, companies should conduct full audits. Companies that disclose personal data to third parties should at a minimum ensure that the recipients are under enforceable contractual obligations to adhere to these principles, unless they are required by law to do otherwise.

# Appendix B: Comparison of the Consumer Privacy Bill of Rights to Other Statements of the Fair Information Practice Principles (FIPPs)

| Consumer Privacy Bill of Rights | OECD Privacy Guidelines (excerpts) | DHS Privacy Policy (generalized) | APEC Principles (excerpts) |
|---|---|---|---|
| **Individual Control.** Consumers have a right to exercise control over what personal data that companies collect from them and how they use it. | **Use Limitation Principle.** Personal data should not be disclosed ... except "with the consent of the data subject or by the authority of law." | **Individual Participation.** Organizations should involve the individual in the process of using PII [personally identifiable information] and, to the extent practicable, seek individual consent for the collection, use, dissemination, and maintenance of PII. | **Choice.** Where appropriate, individuals should be provided with clear, prominent, easily understandable, accessible and affordable mechanisms to exercise choice in relation to the collection, use and disclosure of their personal information. |
| **Transparency.** Consumers have a right to easily understandable information about privacy and security practices. | **Openness Principle.** There should be a general policy of openness about developments, practices and policies with respect to personal data. | **Transparency.** Organizations should be transparent and notify individuals regarding collection, use, dissemination, and maintenance of PII. | **Notice.** Personal information controllers should provide clear and easily accessible statements about their practices and policies .... |

| Consumer Privacy Bill of Rights | OECD Privacy Guidelines (excerpts) | DHS Privacy Policy (generalized) | APEC Principles (excerpts) |
|---|---|---|---|
| **Respect for Context.** Consumers have a right to expect that companies will collect, use, and disclose personal data in ways that are consistent with the context in which consumers provide the data. | **Purpose Specification Principle.** The purposes for which personal data are collected should be specified not later than at the time of data collection and the subsequent use limited to the fulfillment of those purposes or such others as are not incompatible with those purposes and as are specified on each occasion of change of purpose. | **Purpose Specification.** Organizations should specifically articulate the authority that permits the collection of PII and specifically articulate the purpose or purposes for which the PII is intended to be used. | **Notice.** All reasonably practicable steps shall be taken to ensure that such notice is provided either before or at the time of collection of personal information. Otherwise, such notice should be provided as soon after as is practicable. |
|  | **Use Limitation Principle.** Personal data should not be disclosed, made available or otherwise used for purposes other than those specified in accordance with Paragraph 9 [purpose specification] except . . .<br><br>(a) with the consent of the data subject; or<br><br>(b) by the authority of law. | **Use Limitation.** Organizations should use PII solely for the purpose(s) specified in the notice. Sharing PII should be for a purpose compatible with the purpose for which the PII was collected. | **Uses of Personal Information.** Personal information collected should be used only to fulfill the purposes of collection and other compatible or related purposes except: a) with the consent of the individual whose personal information is collected; b) when necessary to provide a service or product requested by the individual; or, c) by the authority of law and other legal instruments, proclamations and pronouncements of legal effect. |
| **Security.** Consumers have a right to secure and responsible handling of personal data. | **Security Safeguards Principle.** Personal data should be protected by reasonable security safeguards against such risks as loss or unauthorized access, destruction, use, modification or disclosure of data. | **Security.** Organizations should protect PII (in all media) through appropriate security safeguards against risks such as loss, unauthorized access or use, destruction, modification, or unintended or inappropriate disclosure. | **Security Safeguards.** Personal information controllers should protect personal information that they hold with appropriate safeguards against risks, such as loss or unauthorized access to personal information, or unauthorized destruction, use, modification or disclosure of information or other misuses. |

| Consumer Privacy Bill of Rights | OECD Privacy Guidelines (excerpts) | DHS Privacy Policy (generalized) | APEC Principles (excerpts) |
|---|---|---|---|
| **Access and Accuracy.** Consumers have a right to access and correct personal data in usable formats, in a manner that is appropriate to the sensitivity of the data and the risk of adverse consequences to consumers if the data is inaccurate. | **Individual Participation Principle.** An individual should have the right: a) to obtain from a data controller, or otherwise, confirmation of whether or not the data controller has data relating to him; b) to have communicated to him, data relating to him within a reasonable time; at a charge, if any, that is not excessive; in a reasonable manner; and in a form that is readily intelligible to him; c) to be given reasons if a request made under subparagraphs(a) and (b) is denied, and to be able to challenge such denial; d) to challenge data relating to him and, if the challenge is successful to have the data erased, rectified, completed or amended. | **Data Quality and Integrity.** Organizations should, to the extent practicable, ensure that PII is accurate, relevant, timely, and complete. | **Access and Correction.** Individuals should be able to: a) obtain from the personal information controller confirmation of whether or not the personal information controller holds personal information about them; b) have communicated to them, after having provided sufficient proof of their identity, personal information about them; i. within a reasonable time ii. at a charge, if any, that is not excessive; iii. in a reasonable manner; iv. in a form that is generally understandable; and, c) challenge the accuracy of information relating to them and, if possible and as appropriate, have the information rectified, completed, amended or deleted. |
| | **Data Quality Principle.** Personal data should be relevant to the purposes for which they are to be used, and, to the extent necessary for those purposes, should be accurate, complete and kept up-to-date. | | **Integrity of Personal Information.** Personal information should be accurate, complete and kept up-to-date to the extent necessary for the purposes of use. |
| | | | **Preventing Harm.** Recognizing the interests of the individual to legitimate expectations of privacy, personal information protection should be designed to prevent the misuse of such information. |

| Consumer Privacy Bill of Rights | OECD Privacy Guidelines (excerpts) | DHS Privacy Policy (generalized) | APEC Principles (excerpts) |
|---|---|---|---|
| **Focused Collection:** Consumers have a right to reasonable limits on the personal data that companies collect and retain. | **Collection Limitation Principle.** There should be limits to the collection of personal data and any such data should be obtained by lawful and fair means and, where appropriate, with the knowledge or consent of the data subject. | **Data Minimization:** Organizations should only collect PII that is directly relevant and necessary to accomplish the specified purpose(s) and only retain PII for as long as is necessary to fulfill the specified purpose(s). | **Collection Limitation.** The collection of personal information should be limited to information that is relevant to the purposes of collection and any such information should be obtained by lawful and fair means, and where appropriate, with notice to, or consent of, the individual concerned. |
| **Accountability.** Consumers have a right to have personal data handled by companies with appropriate measures in place to assure they adhere to the Consumer Privacy Bill of Rights. | **Accountability Principle.** A data controller should be accountable for complying with measures which give effect to the principles stated above. | **Accountability and Auditing:** Organizations should be accountable for complying with these principles, providing training to all employees and contractors who use PII, and auditing the actual use of PII to demonstrate compliance with these principles and all applicable privacy protection requirements. | **Accountability.** A personal information controller should be accountable for complying with measures that give effect to the Principles stated above. When personal information is to be transferred to another person or organization, whether domestically or internationally, the personal information controller should obtain the consent of the individual or exercise due diligence and take reasonable steps to ensure that the recipient person or organization will protect the information consistently with these Principles. |

www.ingramcontent.com/pod-product-compliance
Lightning Source LLC
Chambersburg PA
CBHW081225170526
45165CB00009B/2953